Advanced Sciences and Technologies for Security Applications

Indexed by SCOPUS

The series Advanced Sciences and Technologies for Security Applications comprises interdisciplinary research covering the theory, foundations and domain-specific topics pertaining to security. Publications within the series are peer-reviewed monographs and edited works in the areas of:

- biological and chemical threat recognition and detection (e.g., biosensors, aerosols, forensics)
- crisis and disaster management
- terrorism
- cyber security and secure information systems (e.g., encryption, optical and photonic systems)
- traditional and non-traditional security
- energy, food and resource security
- economic security and securitization (including associated infrastructures)
- transnational crime
- human security and health security
- social, political and psychological aspects of security
- recognition and identification (e.g., optical imaging, biometrics, authentication and verification)
- smart surveillance systems
- applications of theoretical frameworks and methodologies (e.g., grounded theory, complexity, network sciences, modelling and simulation)

Together, the high-quality contributions to this series provide a cross-disciplinary overview of forefront research endeavours aiming to make the world a safer place.

The editors encourage prospective authors to correspond with them in advance of submitting a manuscript. Submission of manuscripts should be made to the Editor-in-Chief or one of the Editors.

More information about this series at http://www.springer.com/series/5540

Eva Moehlecke de Baseggio
Olivia Schneider • Tibor Szvircsev Tresch
Editors

Social Media and the Armed Forces

Springer

ERGOMAS

Editors
Eva Moehlecke de Baseggio
Military Academy (MILAC) at ETH Zürich
Birmensdorf, Switzerland

Olivia Schneider
Military Academy (MILAC) at ETH Zürich
Birmensdorf, Switzerland

Tibor Szvircsev Tresch
Military Academy (MILAC) at ETH Zürich
Birmensdorf, Switzerland

ISSN 1613-5113 ISSN 2363-9466 (electronic)
Advanced Sciences and Technologies for Security Applications
ISBN 978-3-030-47510-9 ISBN 978-3-030-47511-6 (eBook)
https://doi.org/10.1007/978-3-030-47511-6

This Springer imprint is published by the registered company Springer Nature Switzerland AG
The registered company address is: Gewerbestrasse 11, 6330 Cham, Switzerland

Preface

More than 30 years ago, at the age of twenty, I joined the Swiss Armed Forces as one of around 35,000 new recruits. Until today, this day remains memorable to me as it marked the transition from a young man's normal everyday life to being part of a uniform collective, subjected to a rather strict set of rules. The thing that struck me the most was how at times I felt completely cut off from my previous environment. Back then, I did not have a smartphone on which to look at the Instagram pictures of my friends. Neither was I able to post pictures of the blisters I got during the first march to relieve some tension, or even to look up simple tricks for how to treat them. In addition, I did not have a fitness tracker that could reveal the location where I was to do my workout.

The change from my civil life to becoming a soldier was overwhelming and all-encompassing. Nevertheless, I managed to find my way. After finishing my academic education more than two decades ago in 1999, I started my academic work in the Defence sector. In 2008, I became Head of Military Sociology at the Military Academy (MILAC) at the ETH Zurich in Switzerland. Ever since the start of my academic career, I have been an active member of the 'European Research Group on Military and Society' (ERGOMAS) and the Research Committee 'Armed Forces and Conflict Resolution' of the International Sociological Association (ISA RC01). Furthermore, I continued my military service as a special officer in the rank of captain until the end of 2014. This has enabled me to observe the constant changes the armed forces are undergoing on both the national and the international level, one of them being the changes related to new communication technologies.

Within the last 20 years, the world we live in has undergone profound transformations. The advancement of the World Wide Web, the dynamic and interactive Web 2.0, was designed for collaboration and participation. Web 2.0 enables users to communicate almost in real time and to combine and exchange dynamic information online, thus reducing the need to physically exchange document-based information to a minimum. Social media is a key player amongst the communication tools based on Web 2.0. Its use has been particularly impacted by the triumphant market launch of smartphones in 2007.

Just as letterpress printing was not only a technological innovation but also changed society as a whole, Web 2.0 and social media affect the way we think, behave, and communicate. At the dawn of the 2020s, social media has – in line with the great sociologist Émile Durkheim – become a social fact. It is an integral element of everyday life for large parts of the world's population and takes concepts like the information society, the knowledge society or the network society to a new level, hence the interest of many scholars.

As a social fact, social media does not stop at the doorstep of the Defence sector. Armed forces, whatever system they may be based on (professional, conscript or militia), above all consist of people. These people are not only members of the armed forces, they also have a life before, alongside, and after their service. Most of them are young and belong to the cohorts of the digital natives and the always-on-generation. It goes without saying that they will not put aside their smartphones once they join the armed forces, and neither will they cease using social media. What does this mean for the armed forces?

Armed forces are organisations characterised by unity and secrecy. They usually communicate very formally and factually – both internally and externally. Social media, in turn, stands for open, informal, and emotional communication. This potentially leads to tensions between social media and armed forces, which can result in a range of new threat scenarios for the latter. Social media thus holds new forms of vulnerability for the armed forces, the state, and society. Sensitive information published by military personnel on social media – wilfully or not –, misinformation or disinformation scattered on social media, or targeted hacker attacks via social media represent unprecedented dangers and risks with which the armed forces have to deal. These are opposed to the many benefits and chances that this new way of communication brings. Social media opens up promising opportunities for civil-military relations. Although traditional mass media continue to play an important role in how society's ideas about the armed forces are constructed, social media enables the armed forces to reach a potentially vast audience on direct channels, without a gatekeeper in between. In addition, social media offers the possibility of interaction. The armed forces no longer have to rely on one-way communication, but can interact with their community and get first-hand impressions of how people feel about them.

This book examines the phenomenon of social media from a sociological perspective, aiming to point out current developments, potentials, and risks for armed forces throughout the world. It comprises 12 contributions by 28 scholars from 10 countries, mostly members of either the 'European Research Group on Military and Society' (ERGOMAS) or the Research Committee 'Armed Forces and Conflict Resolution' of the International Sociological Association (ISA). In addition to some very renowned authors with longstanding expertise and a rich fundus of experience, the book features young scholars who probably look at the subject differently. I consider this mix of experienced and young scholars a great advantage. On the one hand, it includes the voices of those who might have the experience and knowledge to critically assess new developments. On the other hand, the research field of social media is still young, and as such does not necessarily require years of experience, but can actually profit from fresh perspectives. Young scholars might even understand the phenomenon of social media better, as they started living with it at an earlier age.

At any rate, I hope the current book will support armed forces and military scholars around the globe in dealing with the different social media platforms and the challenges they present – because one thing has become clear within the last decade: social media is here to stay.

Zurich, Switzerland Tibor Szvircsev Tresch
2020

Contents

About the Editors and Contributors

Editors

Eva Moehlecke de Baseggio manages the research project 'Social Media as a Communication Channel of the Swiss Armed Forces' in the Department of Military Sociology at the Military Academy (MILAC) at the ETH Zurich, Switzerland, where she has been working since 2016. She analyses the effects of the social media communication of the Swiss Armed Forces on their legitimacy in Swiss society and on the motivation of youth to join the military. Her research interests lie in public opinion formation, digital sociology, and social legitimacy processes. Eva holds a Master of Arts in Sociology and Cultural Sciences from the University of Zurich. eva.moehlecke@moehlecke.com

Olivia Schneider is currently doing her PhD in Sociology at the University of Zurich. She researches the online social negotiation processes concerning the Swiss Armed Forces. Since 2017, Olivia has been a research assistant in the Department of Military Sociology at the Military Academy (MILAC) at the ETH Zurich. She works on the project 'Social Media as a communication channel of the Swiss Armed Forces'. In this context, Olivia has contributed several times to the annually published study series 'Sicherheit'. She also heads a sub-project that analyses the content of communication by the Swiss Armed Forces and related discussions on social media. oliviaschneider@gmx.net

Tibor Szvircsev Tresch (PhD) studied Sociology, Political Science, and Criminology at the University of Zurich, receiving his doctorate in 2005. Since 2008 he has been a lecturer in Military Sociology at the Military Academy (MILAC) at the ETH Zurich. In 2008, he became the coordinator of the Working Group on Recruitment and Retention in the 'European Research Group on Military and Society' (ERGOMAS). Since 2014, he has been vice president of Research

Committee 01 of the International Sociological Association (ISA RC01). From August 2018 to July 2019, he was a Visiting Researcher at the Saltzman Institute of War and Peace Studies (SIWPS) at Columbia University, New York. tibor.szvircsev@vtg.admin.ch

Contributors

John Bornmann (PhD) received his PhD from George Washington University in 2009, writing on the experiences of new soldiers going through basic training. While completing his dissertation, he was deployed in support of the 345th Psychological Operations Company to Baghdad in 2004 and returned to Iraq twice in a civilian role supporting MNF-I and USF-I in conducting surveys and interviews with local Iraqi residents. He has been working at MITRE since 2010 as a qualitative data analyst and military experience SME. jbornmann@mitre.org

Karin K. De Angelis (PhD) is an associate professor of Sociology in the Department of Behavioral Sciences & Leadership at the US Air Force Academy (USAFA). She joined the USAFA faculty in 2011 and serves as the department's socio-cultural discipline lead. Her research interests include race/ethnicity in the US military, with a focus on Hispanic servicemembers; the intersection of gender, work, and family; military families; sexual assault prevention; and diversity in organisations. karin.deangelis@usafa.edu

Sara Beth Elson (PhD) is a Behavioural Scientist with The MITRE Corporation. She studies the ways in which people use technology and the implications of that usage for well-being, achievement, political activism, and other topics. In particular, she has studied the implications of social media use within diverse populations, including the military and populations experiencing political turmoil. She has also led the development of new techniques for analysing social media, with an emphasis on indicators of emotion and how changes in these indicators can be analysed mathematically. She holds a PhD in Social Psychology from Ohio State University. selson@mitre.org

Morten G. Ender (PhD) is Professor of Sociology at the United States Military Academy at West Point, New York, USA. He has taught at the University of Maryland European Division, the University of Maryland; Marymount University; the University of North Dakota; the American College of Norway; Mount Saint Mary's College; and the Universität der Bundeswehr München. He writes about the military, war, and teaching. His articles have been published in *Teaching Sociology*, *Journal of Adolescence*, *Death Studies*, *Military Psychology*, *Journal of Homosexuality*, *War & Society*, and *Armed Forces & Society*. He is the Book Review Editor for *Res Militaris: The European Journal of Military Studies*. morten.ender@westpoint.edu

Nicola T. Fear (DPhil) joined the Academic Department of Military Mental Health at King's College London in 2004, having trained as an epidemiologist at the London School of Hygiene and Tropical Medicine and at the University of Oxford. She has also worked as an epidemiologist within the UK Ministry of Defence. Since 2011, she has been Director of the King's Centre of Military Health Research alongside Professor Sir Simon Wessely. In 2014, Nicola was awarded a Chair in Epidemiology. She is the lead epidemiologist on the KCMHR's military cohort study and heads several studies examining the impact of military service on families. nicola.t.fear@kcl.ac.uk

Sofia Martins Geraldes is a PhD candidate in History, Studies of Security and Defense (ISCTE-IUL & Military Academy), and holds a PhD scholarship. Her thesis is about the impact of external online disinformation, disseminated through social media, on the security governance of the EU and NATO. She was the founder and coordinator of EuroDefense-Jovem and worked as trainee at the National Defence Institute. She holds a BA in Political Science and International Relations and an MA in International Relations (University of Beira Interior). Her research interests include international relations theory, security studies, influence operations, cyberspace, the EU, and NATO. sofia_cristina_geraldes@iscte_iul.pt

Rachael Gribble (PhD) is a Lecturer in War and Psychiatry at King's College London. With a background in public health, her work focuses on military families, women's health, and public attitudes towards the military, using a mixture of quantitative and qualitative methodologies. rachael.gribble@kcl.ac.uk

Arita Holmberg (PhD, Stockholm University, Sweden) is an associate professor and senior lecturer in Political Science in the Department of Security, Strategy and Leadership at the Swedish Defence University. She has published within the field of security and defence transformation, gender and the military, and security and defence policy. Her current research concerns social and political challenges for the military organisation as well as children, security, and sustainability. Her recent articles have been published, among others, in *Defence Studies*, *Childhood*, and *Gender, Work and Organization*. arita.holmberg@fhs.se

Aki-Mauri Huhtinen (LTC (GS), PhD) is a military professor in the Department of Leadership and Military Pedagogy at the Finnish National Defence University. His areas of expertise are military leadership, command and control, information warfare, and the philosophy of science in military organisational research. He has published peer-reviewed journal articles and books on information warfare and non-kinetic influence. aki.huhtinen@mil.fi

Jutta Joachim (PhD) is a senior lecturer of International Relations at Radboud University. She received her PhD in Political Science from the University of Wisconsin-Madison and her MA in International Studies from the University of South Carolina. She is the author of *Agenda Setting, the UN, and NGOs: Gender*

Violence and Reproductive Rights (Georgetown University Press 2007), co-author of *Private Security and Identity Politics: Ethical Hero Warriors, Professional Managers and New Humanitarians* (Routledge 2018), and co-editor of *International Organizations and Implementation: Enforcers, Managers, Authorities* (Routledge 2007) and *Transnational Activism in the UN and the EU: A Comparative Study* (Routledge 2008). Numerous articles of hers have appeared in international peer-reviewed journals and edited volumes. j.joachim@fm.ru.nl

Jelena Juvan (PhD) is an assistant professor in the Department of Defence Studies at the Faculty of Social Sciences, University of Ljubljana, and a senior research fellow at the Defence Research Centre at the aforementioned Faculty. She is an assistant professor of the following courses: Common Foreign and Security Policy of the European Union, Common Security and Defence Policy of the European Union, Disaster Management, Informatisation and Contemporary Armed Forces, Defence and Security System, and Security in Information Society. As a researcher, she has been involved in several national and international research projects. She has been acting as an independent evaluator for the EU FP7 and H2020 Research Projects since 2007. From 2013 to 2015, she was a member of the NATO International Scientific Evaluation Group (NATO ISEG 2013–2015). Her research fields include European security, crisis management, contemporary armed forces, use of digital media, and ICT in contemporary armed forces. jelena.juvan@fdv.uni-lj.si

Ryan Kelty (PhD) is an associate professor of Sociology at the US Air Force Academy. He is an award-winning teacher and scholar whose research focuses on diversity and social psychology issues in the military. Dr Kelty has consulted with numerous national and international defence organisations. He is the co-editor of *Private Military and Security Contractors: Controlling the Corporate Warrior* (Rowman & Littlefield, 2016) and *Risk-Taking in Higher Education: The Importance of Negotiating Intellectual Challenge in the College Classroom* (Rowman & Littlefield, 2017). He earned his BA from Middlebury College and his PhD from the University of Maryland. Dr Kelty previously served on the faculty at West Point and Washington College. ryan.kelty@usafa.edu

Gerhard Kümmel (PhD) is a senior researcher at the Center for Military History and Social Sciences of the Bundeswehr in Potsdam, Germany. He studied Political Science, Sociology, and History at the Philipps-University in Marburg, Germany, and received the J. William Fulbright Dissertation Award in 1996. From 2010 to 2014, he was the president of the Research Committee 01: Armed Forces and Conflict Resolution (RC01) of the International Sociological Association (ISA). His work centres on women in the military, the representation of the military in film, and veterans' issues. gerhardkuemmel@bundeswehr.org

Daniel Leightley (PhD) is a Post-Doctoral Research Associate at the King's Centre for Military Health Research (KCMHR). His research focuses on the interface between machine learning and mobile health technologies, specifically on the diagnosis, treatment, intervention, and management of physical and mental health conditions in the British Armed Forces community. daniel.leightley@kcl.ac.uk

Michael D. Matthews (PhD) is Professor of Engineering Psychology at the United States Military Academy. He is a Templeton Foundation Senior Positive Psychology Fellow, a Fellow of the Army Chief of Staff's Strategic Studies Group (2014–2015), and the author of *Head Strong: How Psychology is Revolutionizing War* (Oxford University Press, 2014) as well as numerous other books, chapters, and scientific articles. Dr Matthews is a founding member of the Military Child Education Coalition's Science Advisory Board and a leadership consultant to governmental agencies, professional sports teams, and other organisations. mike.matthews@westpoint.edu

Teija Norri-Sederholm (PhD) is an adjunct professor in the Department of Leadership and Military Pedagogy at the Finnish National Defence University. Her main research areas are situational awareness, inter-organisational communication in command centres and hybrid environments, national security, and the dark side of social media. teija.norri-sederholm@mil.fi

Elisa Norvanto currently works at the Ministry of Defence as a Special Adviser in the defence cooperation unit. Her permanent post is at Laurea UAS, where she serves as director of Research & Development. Norvanto holds a master's degree in Political Sciences and is currently doing her PhD at the National Defence University. Her research interests relate to trust in international security and defence cooperation. elisa.norvanto@mil.fi

Keith Paulson is an architect and engineer at MITRE, which operates federally funded research and development centres and is dedicated to solving hard problems for a safer world. Keith has lived and worked as an engineer and director across the Middle East, Far East, Europe, and the USA. He graduated with dual degrees in Electrical Engineering and Computer Engineering from the University of Missouri and a Master's degree in Software Engineering from the University of Colorado. He actively works in system design, artificial intelligence, and social impact of technology. kpaulson@mitre.org

Andrea Rinaldo graduated from the University of Zurich in 2014 with a Master of Arts in Sociology. She has been working as a research assistant at the Military Academy (MILAC) at the ETH Zurich since 2014. Her research interests concentrate on diversity and inclusion in organisations, military sociology, public opinion, and qualitative research methods. She is currently working on her PhD on diversity management in the Swiss Armed Forces. The thesis focuses on experiences of minorities in the armed forces, their perception of inclusion, as well as on inclusion strategies of military organisations. rinaldoandrea@gmx.ch

Shira Rivnai Bahir is a Digital Sociology researcher. She is a PhD Candidate in the Department of Sociology and Anthropology, Ben-Gurion University of the

Negev (Negev and Darom Scholarship). Her doctoral research explores the role of anonymity and secrecy practices in digital activism. In addition, Shira is a researcher at the IDF Behavioral Science Center. Her studies focus on civil-military relations, digital activism, and the sociology of AI. songa7@gmail.com

David E. Rohall (PhD) is Professor of Sociology and Department Head of the Department of Sociology & Anthropology at Missouri State University. He has been studying trends in American society generally and the military services specifically for almost 20 years. His most recent works include Symbolic Interaction in Society (Rowman & Littlefield, 2019) and Inclusion in the American Military: A Force for Diversity (Lexington Press, 2017). drohall@missouristate.edu

Andrea Schneiker (PhD) is a junior professor of Political Science at the University of Siegen. She received her PhD in Political Science from the University of Münster, after which she worked as an assistant professor at the Leibniz University of Hannover and the University of Bremen. She is the author of *Humanitarian NGOs, (In)Security and Identity* (Routledge, 2015) and co-editor of *Researching Non-state Actors in International Security* (with Andreas Kruck; Routledge, 2017). Her articles have been published in numerous journals, including *International Studies Review*, *International Studies Perspectives*, *Comparative European Politics*, *Millennium*, *Security Dialogue*, *Cambridge Review of International Affairs*, *Contemporary Security Policy*, *European Policy*, and *Global Policy*. andrea.schneiker@googlemail.com

Marie-Louise Sharp (PhD) is a Senior Research Associate at the King's Centre for Military Health Research. She is a mixed methods researcher with research expertise in psychological medicine and epidemiology and is responsible for KCMHR research strategy. Her research interests include military mental health and help-seeking behaviours as well as the health and well-being of Emergency Responders and focuses on methods to extend research impact. Marie-Louise previously worked in the non-profit sector, specialising in armed forces healthcare policy and advising the UK Government on research and best practice. marie-louise.sharp@kcl.ac.uk

Uroš Svete (PhD) is an associate professor in the field of Defence Science at the Faculty of Social Sciences, University of Ljubljana. From 2011 to 2015, he was Chair of the Department of Defence Studies at the Faculty of Social Sciences. His research interest spans areas such as information technology and security issues, analysis of contemporary (asymmetric) conflicts, military/defence technology, security theories, and security concepts in contemporary societies. uros.svete@fdv.uni-lj.si

Karoliina Talvitie-Lamberg (PhD) is a post-doctoral research fellow in the Faculty of Information Technology and Communication Sciences at the University of Tampere, specialising on participation in social media and video-streaming cultures as well as the formation of the digital social self. Recent research projects include datafication and vulnerability, AI for social inclusion, social bots, DIY surveillance in self-care technologies, and data-driven AI solutions for health and well-being. karoliina.talvitie-lamberg@tuni.fi

Victoria Williamson (PhD) is a Post-Doctoral Research Associate at the King's Centre for Military Health Research. Her research focuses on military-related trauma, moral injury, and Post-Traumatic Stress Disorder. victoria.williamson@kcl.ac.uk

Introduction

Eva Moehlecke de Baseggio and Olivia Schneider

Abstract Social media has fundamentally changed communication and interaction in today's societies over the past two decades. This applies not only to the individual members of a society, but also to its institutions and organisations. The tendencies of isomorphism, which is the process of adopting processes and strategies that have proven to be successful in other organisations (DiMaggio PJ, Powell WW: Am Sociol Rev 48(2):147–160, 1983), force public sector organisations such as armed forces to establish social media as part of their reality. Yet, what is it that makes social media the phenomenon it is?

Keywords Social media in the everyday life of military personnel · Gender-specific representation on social media in the military · Social media discussions as insights into public opinion · Social media in the military: risks and dangers

Social media has fundamentally changed communication and interaction in today's societies over the past two decades. This applies not only to the individual members of a society, but also to its institutions and organisations. The tendencies of isomorphism, which is the process of adopting processes and strategies that have proven to be successful in other organisations (DiMaggio and Powell 1983), force public sector organisations such as armed forces to establish social media as part of their reality. Yet, what is it that makes social media the phenomenon it is?

Essentially, social media platforms are communication channels that allow users to share information, including audio and visual content such as pictures and videos. On social media, information is conveyed and exchanged in almost real-time. Users can connect many-to-many, which in itself is already an innovation compared to other communication channels. Social media in its nature is about collaboration and participation, with the intent to bring people together. People communicate in an informal, open, and ideally deliberate way. Thus, as is evident in the emergence

E. Moehlecke de Baseggio (✉) · O. Schneider (✉)
Military Academy (MILAC) at ETH Zurich, Birmensdorf, Switzerland
e-mail: eva.moehlecke@moehlecke.com; oliviaschneider@gmx.net

E. Moehlecke de Baseggio et al. (eds.), *Social Media and the Armed Forces*, Advanced Sciences and Technologies for Security Applications, https://doi.org/10.1007/978-3-030-47511-6_1

of new forms of communitisation, for instance Facebook groups or family chats, social media is a reflection of societal changes and, at the same time, changes society itself (Hinton and Hjorth 2012). These platforms are not mere media, but have become institutions in the sociological sense, as they structure the way people behave and relate to each other. Social media as an institution sets societal norms, such as the handling of private data and privacy in general. Especially younger people are fully immersed in these platforms, and are consequently adapting their communication and information behaviour completely. This can be attributed to the fact that social media particularly serves the preferences of young people, who favour pictures and videos over purely textual media. Many studies show the displacement of more traditional communication and information channels in favour of social media platforms by young people (Fög 2018, 2019; Reuters Institute 2019).

Building on the media aspect, there is another relevant novelty compared to more traditional communication channels. In times of newspapers and television or radio broadcasting, journalists traditionally were the gatekeepers of information. On social media, however, anyone can communicate – private persons, companies, public sector agencies, and media outlets. On the one hand, this is a democratising function of social media, whereas, on the other hand, it fosters the proliferation of fake news and misinformation. Furthermore, not only can anyone communicate, but the contents can also be debated anytime, anywhere, and by anyone. While this has, of course, always been possible orally, with social media the formerly face-to-face conversations about media content have become written and therefore enduring (Katzenbach 2017). This is a change which should not be underestimated. The textuality of these new forms of conversations influences their endurance and hence their accessibility as well as their consequences, which is illustrated by convictions for hate speech on the web, for example.

Armed forces rely on societal acceptance; as cost-intensive organisations they need legitimacy in order to exist. Trust is a precursor of legitimacy (Suchmann 1995). Thus, for people to invest trust in them, armed forces need to communicate in an open, transparent, competent, ethical, and adequate manner – and with the appropriate frequency (Bentele 1994). This fosters the creation of public trust (ibid.). In this context, social media becomes an important communication channel for armed forces. It is especially adequate for the main target group of armed forces, namely youth, and furthermore serves to inform society about them and to increase their presence. Besides that, armed forces also need to grapple with social media because their recruits, cadets, soldiers, and officers are human beings not excluded from the comprehensive phenomenon of social media. Thus, armed forces need to understand the everyday implications of social media, and to build up awareness towards the strategic chances, opportunities, and threats social media brings, both generally and specifically, for armed forces.

In this anthology, the opportunities and risks with which various Western armed forces feel confronted in relation to social media, and which therefore are part of military sociological research, are collected and highlighted. This book is the first work to bundle the state of research on social media usage in armed forces. Thus, it not only offers fascinating insights into the handling of social media for a (military) academic audience interested in findings from the scope of communication, organisational sociology, recruitment, diversity, armed forces, and democratic control, but is also of particular importance for political decision-makers, defence practitioners, and other public sector organisations. The way in which the various challenges social media poses to democracies and armed forces are handled is examined from an international perspective and scientifically processed, and guidance is provided on how to deal with this type of media. This work is intended to provide an overview of the state of military social-scientific research on the subject of social media and armed forces, highlighting both the positive and the negative aspects. Four topics emerged in the process, which in the following are portrayed briefly.

1 Social Media in the Everyday Life of Military Personnel

As is the case for all other spheres of life, digitalisation has also permeated the armed forces. The distinction between offline reality and online virtual existence has become obsolete, as is expressed by terms like 'onlife' or the 'digital mundane' (Floridi 2014; Maltby and Thornham 2016). Digital technology and gadgets are completely interwoven in today's everyday life, shaping the daily routines of people, which is illustrated by habits such as checking for free Wi-Fi or taking selfies (Maltby and Thornham 2016). The example of the Swiss Armed Forces sports app 'ready #teamarmee', which prepares future recruits for the physical requirements of military service, shows that this has become a reality for armed forces, as well.

As part of these developments, social media has also become part of the ordinary infrastructure we use, becoming visible through its everyday use and reaching normative character, for instance in tying success to visibility (Maltby and Thornham 2016). Digitalisation in the armed forces includes the everyday life of the individual recruits, cadets, soldiers and officers, too. In times of the internet, social media, and smartphones, soldiers are potentially permanently available and only one click away. Military personnel use social media just as any other population group does. An analysis of the tweets of military personnel shows that their tweets are about their everyday life in the armed forces, including the use of mundane military slang (Pavalanathan et al. 2016). Hence, soldiers making their doings visible can be understood as transferring societal norms into the realm of the armed forces. Different to other societal groups, however, the business of military personnel is the security of the respective nation and its people. This raises questions about a possible influence of social media on the effectiveness of armed forces, be it as a distraction only or in a more aggravated form. Are there differences between the social media usage of soldiers and officers and civilians, and does this affect performance of armed forces?

In the first chapter of this anthology, Karin K. De Angelis, Ryan Kelty, Morten G. Ender, David E. Rohall, and Michael D. Matthews elaborate how comprehensively the phenomenon of social media influences the lives of young people in particular, including their education. In their contribution, *Ubiquity with a Dark Side:*

Civil-Military Gaps in Social Media Usage, they pursue the question whether there is a difference in the social-media related behaviour of civilian students and military cadets. The results of their research suggest a gap between the groups, with civilian students spending more time on social media and suffering from greater distraction both in class and when doing homework. Military cadets, on the other hand, seem to be less exposed to social media and therefore have less contact with the dark side of social media. The research thus indicates differences in the everyday social media behaviour of civilian students and military cadets.

In the second contribution, *Social Media Use at a U.S. Military Academy: Perceived Implications for Performance and Behavior*, Sara Beth Elson, Ryan Kelty, Keith Paulson, John Bornmann, and Karin K. De Angelis explore a similar path. They focus on the impact of social media on the academic performance of military cadets, surveying both cadets and faculty members. Just as their civilian counterparts, cadets spend too much time on social media, whereby self-assessment and external assessment by faculty members correspond. Cadets and faculty members agree that this impacts academic performance negatively. In its positive manifestation, however, social media enables cadets to break through the isolation of their education at the academy and keep in touch with their families, friends, and partners. A surprising result of this research is the attractiveness and popularity of a social media platform which is considered much less popular among civilian students. Jodel, a social media platform that grants anonymity, seems to serve the specific needs of cadets seeking an outlet to express opinions that are not in line with the values and attitudes of military culture.

The everyday use of social media does not only relate to the individual members of armed forces. Armed forces as an organisation also use social media as communication channels. This organisational use of social media exerts influence on military personnel just as much as does individual use. Eva Moehlecke de Baseggio illustrates this in her contribution, *The Need for Visibility: The Influence of Social Media Communication on the Swiss Armed Forces' Officers*. The results of semi-structured interviews with cadre members of the Swiss Armed Forces suggest that military personnel feel that society is not informed well enough about the armed forces. They link this lack of knowledge to a certain disregard for the organisation and its members. Conversely, they connect the external visibility of the organisation to the perceived appreciation for what they do, which translates further into the degree of organisational identification and affective commitment they feel. Cadre members in particular stress the importance of social media platforms for young people, who are the future recruits of the Swiss Armed Forces.

2 Gender-Specific Representation on Social Media

Social media platforms are an ideal tool to make use of the power of rich imagery, which usually is something that armed forces dispose of in abundance. Research proves that visual media evoke more emotions in the audience (Fahmy et al. 2006).

Emotions, in turn, attract more attention (Hendricks and Vestergaard 2019), hence armed forces' communication will benefit from this. Simultaneously, however, another process takes place. The distinct visual character of social media and its abundance fosters the process of social learning (Yılmaz et al. 2019). Social learning theory goes back to Albert Bandura (1977, 2001), stating that people internalise what they perceive through the media and learn from it by deriving consequences for their own actions and attitudes. The main target group of armed forces' social media communication, youth, will consequently be the group most affected by visual content. This happens at a time in the lives of young people when the role of the media as an instance of socialisation is of great importance (Mühler 2008). The media convey models and guiding principles, which, among other things, impact the formation of identity of young people (ibid.). In this regard, gender roles are relevant elements. In fact, the media rank among the most powerful and omnipresent influences on the perception of gender roles (Wood 1994).

As armed forces are traditionally regarded as gendered organisations that transport a stereotypical image of masculinity, opening them to women potentially challenges this stereotype and the idea of the masculine soldier implicitly contained therein. The question arises which effect the integration of women into armed forces has on gender images. Thus, in their contribution, *Managing Femininity through Visual Embodiment: The Portrayal of Women on the Instagram Accounts of the Swedish and the Swiss Armed Forces*, Andrea Rinaldo and Arita Holmberg compare how the two countries' female soldiers are represented. They find that the ways female soldiers are portrayed reflect both the societal context and the respective characteristics of the respective armed forces. However, this always occurs in relation to the reference category male, thus reproducing the gendered organisation by making use of the power of visuals inherent in Instagram.

At the other end of the spectrum of a military career are veterans. Unlike recruits, they have already undergone the process of adult socialisation by joining the armed forces. Thus, their concept of gender roles was not only affected by their primary and secondary socialisation, but also by tertiary organisational socialisation processes, which, in the case of the military, perpetuate a concept of masculinity that portrays the male military archetype as strong, silent, self-reliant, and loyal (Arkin and Dobrofsky 1978; Caddick et al. 2015). Manliness is expressed through military skills and combat training, and success means to conform to this concept (Arkin and Dobrofsky 1978). It is clear that such a concept of masculinity is problematic especially for wounded veterans. In their contribution, *(Dis-)Empowered Military Masculinities? Recruitment of Veterans by PMSCs through YouTube,* Jutta Joachim and Andrea Schneiker analyse how the YouTube recruitment clips of two U.S. Private Military and Security Companies, or PMSCs, partake in the construction of this stereotypical masculinity. The major recruitment base for PMSCs consists of military veterans, who are explicitly addressed in the corresponding recruitment videos. By performing a content analysis of the recruitment clips, the authors manage to disclose how skilfully PMSCs deconstruct and reassemble the different manifestations of military masculinity by addressing both the skills and needs of veterans, while at the same time restoring them to their place in society and giving them back a sense of self-worth that may have been lost previously. The production of the notion of military masculinity therefore goes beyond armed forces and profits from visual media platforms such as social media.

Thus, the way masculinity and femininity are portrayed in the social media appearances of armed forces potentially influences the identities both of young audiences and of veterans. Armed forces need to consider the responsibility they bear in regard to taking part in shaping societal concepts of gender roles. While modern societies, of course, rely on the reduction of complexity in order to stay functional, we must nevertheless be aware of the reverse side of such complexity reduction, namely prejudices and stereotypes, including stereotyped conceptions of gender roles and abilities. It is not just about social responsibility towards society, youth, and veterans – the representation of gender roles also influences how and which future recruits are addressed.

3 Social Media Discussions as Insights into Public Opinion

The media not only convey gender roles, but also the image a population has of its armed forces. One way to mediate between the military and the civilian is through communication, whereby there naturally are tensions in the media-military relationship (Porch 2002). While bureaucratic organisations like the armed forces tend to do business behind closed doors – especially with regard to war-related issues, the contents of which may be shocking –, the media see themselves as a fourth power whose task it is to monitor the state and thus ensure public control. Military affairs and warfare are political acts about which the people must be informed, and citizens must be convinced that the objectives pursued can be achieved at acceptable costs. The majority of this is done via the mass media, which in turn means that the armed forces are – to a certain degree – dependent on the mass media (Porch 2002). However, traditional mass media are no longer the only major players in the information business. In contrast to established communication media, social media can reach a large audience low-threshold and without a gatekeeper (Jacobs 2016). In addition, social media enables many-to-many communication that has not existed in a comparable form up to now. Thus, it is not only direct organisational communication that is moving into the focus of interest, but also the written and public follow-up discussion of armed forces' social media communities. Online follow-up discussions offer the possibility to interact with citizens, which allows armed forces to reveal themselves to the public and to exercise their control function in a democracy. Moreover, deliberative follow-up discussions on relevant topics enable armed forces to anchor themselves in society through public debate, thus legitimising themselves and their monopoly on the use of force in democracies. The contribution by Olivia Schneider, *The Importance of Discussions on Social Media for the Armed Forces*, explains the importance of such discussions on social media for the armed forces, using a theoretical model that she applies to the communicative situation of armed forces and illustrates with empirical results from reader commentary research.

She thus shows that discussions on social media can be highly significant for armed forces, as they can have a meaningful impact on the image a society has of its armed forces.

The nature of the topics and how they are discussed on social media, however, is often beyond the control of the armed forces, which is contrary to their nature of wanting to keep things under control. Discussions on social media can even create a wave of outrage, a so-called *shitstorm*. Social media content may be picked up by traditional and established media houses, who, on the one hand, need to fulfil their role as fourth estate, monitoring large institutions and addressing possible abuses of power (Porch 2002). On the other hand, the media function according to news values – they want to sell their content and attract the attention of readers. As was elaborated above, the attention of the general public is attracted by emotions (Hendricks and Vestergaard 2019). Scandals can be understood as the climax of emotions and thus draw a maximum of attention. This applies in particular to security policy issues, because these do not often occur in the lives of most citizens (Jacobs 2016). Furthermore, according to Jacobs (2016), the news value of the German Bundeswehr is usually connected with negatively evaluated events, as these evoke more emotions (Hendricks and Vestergaard 2019). In his contribution to this volume, *Debating German Special Forces: A Scandal in the Military, a Documentary, and a Thread*, Gerhard Kümmel shows how a television feature on a scandal in the Bundeswehr, originally produced and broadcast by public television, generates a lot of attention and is the subject of lively discussions on Facebook. In the course of this process of public negotiation of meaning, five topics crystallise that are taken up repeatedly by various commentators. The controversial debate on the feature thus provides an insight into civil-military relations in Germany.

Social media can be understood as a digital gathering in which information is weighed, evaluated, and commented on (Jacobs 2016). Thereby, social media users leave a footprint on social media which allows for insights into the social attitudes and conditions of communities (Pavalanathan et al. 2016). Due to social media, such discussions – which, until recently, were reserved mainly for the private sphere – become accessible and public (Katzenbach 2017). Depending on the issue and platform, users' comments and surrounding data become large data records which allow for new forms of analyses. In their contribution, *Sentiment of Armed Forces Social Media Accounts in the United Kingdom: An Initial Analysis of Twitter*, Daniel Leightley, Marie-Louise Sharp, Victoria Williamson, Nicola T. Fear, and Rachael Gribble research public attitudes towards and perceptions of the British Armed Forces by analysing tweets for linguistic patterns. They find that public perceptions of the British Armed Forces were fairly stable over time, as well as being more positive than negative. However, the time of day plays an important role, as late in the evening and at night, there are more negative tweets than during the day.

Discussions in different groups and forums can be held on various topics, not just on those that professional media players consider relevant according to their mission and news value. Through social media and the internet, like-minded people can be found for virtually any interest or subject one can think of. Thus, alternative topics and innovative patterns of interpretation can be discussed (Schrape 2011).

However, they are also distributed further in society due to their written form and public accessibility. Social media makes it possible to identify subgroups who discuss alternative issues and to analyse the digital exchange of such groups (Pavalanathan et al. 2016). In her contribution, *A Transparent Network – Soldiers' Digital Resistance and Economic Unrest*, Shira Rivnai Bahir selected discussions on social media to find out how certain governmental conditions are discussed by Israeli soldiers and their families. She is able to show how narratives concerning the microeconomic situation of people receive more attention in social media discussions. Through the constant repetition of these narratives by different people in such forums, an individual voice is formed. The topics thus expressed therefore do not correspond to the problems that the mass media address in this context.

4 Risks and Dangers of Social Media

On social media, users do not just leave information about themselves through comments. Social media activities are often linked to other, perhaps unconsciously published data. For instance, a study by Pavalanathan et al. (2016) found that military personnel are more likely to write tweets with geotags than people in the non-military control group. Geotags can reveal sensitive information about the location of troops, especially when deployed abroad, and thus could endanger the mission as well as the soldiers. The danger of social media, especially in a military context, is that sensitive information could end up in the wrong hands (Olsson et al. 2016). In this regard, the heads of the information departments of 28 EU states were asked whether social media could harm the mission of organisations (ibid.). Although most do not see a risk in the national context, many show some concern regarding publishing sensitive information on social media in an international context, while a clear majority sees the greatest danger in the possibility of exposing soldiers abroad to risks by disclosing sensitive information (ibid.). However, social media is not a risk for a state or a society – and thus a challenge for the armed forces – simply due to the disclosure of sensitive information. As was outlined above, mass media shape the image of armed forces and the public attitude towards them (Rukavishnikov and Pugh 2018). Social media must be considered as part of the mass media, but with particularly low access barriers and a wide reach, also internationally. Indeed, social media operations can be launched in one's own country to spread uncertainty and chaos in another country, thus potentially manipulating the recipients. Given that the manipulation of information is not a new tactic, public information has long been an important battlefield which must not be neglected (Porch 2002). However, the situation has been aggravated and changed by social media. Two decades ago, tyrants and terrorists, such as Saddam Hussein, Slobodan Milošević, and Osama bin Laden, discovered mass media as a valuable vehicle for spreading their message (Porch 2002). Moreover, potential enemies no longer need to be in direct contact with established formats, but can share their message directly with the world via social media. Governments and armed forces must therefore also show their point of view. However, caution is called for, because by "manipulating media images for operational advantage, the military courts scepticism and hostility" (Porch 2002, 102).

In her contribution, *The Dark Side of Interconnectivity: Social Media as a Cyber-Weapon,* Sofia Martins Geraldes describes how social media can be used as a weapon. She explains why social media is classified as a cyber-weapon and what dangers it entails. Using the Russia-Ukraine conflict as an example, she illustrates how social media was used to spread fear and insecurity and how this, in turn, can influence people's trust in government and institutions. Furthermore, as part of the population, soldiers are just as much as civilians exposed to mass media communication, which can therefore also have an impact on the attitude and behaviour of soldiers (Rukavishnikov and Pugh 2018).

Similarly, in their contribution, *Misinformation and Disinformation in Social Media as the Pulse of Finnish National Security,* Teija Norri-Sederholm, Elisa Norovanto, Karolina Talvitie-Lamberg, and Aki-Mauri Huhtinen examine the challenges that misinformation and disinformation pose to governments and armed forces in Europe in general and in Finland in particular. On the one hand, democratic governments cannot launch information operations to the same extent as other political systems because of the risk of losing credibility. On the other hand, the question is how to fend off such attacks in order to stabilise society. The authors show that public trust and the provision of true news is a key concept for this. However, measures to ward off mis- and disinformation attacks should not only be implemented nationally, but also in international cooperation, as social media functions transnationally.

The various contributions in this book show that social media is neither just an opportunity nor simply a danger. As a ubiquitous interaction tool, it shapes the lives of prospective soldiers already in training. This potentially results in a higher level of distraction, which in turn – different to prior generations – can be detrimental to their education. Moreover, the anonymity of certain social media platforms promotes cyberbullying. As attacks on comrades and superiors can negatively influence the 'esprit de corps', this could affect the effectiveness of armed forces. An advantage of social media, however, is the fact that the images conveyed through it shape the image of armed forces in society. In addition, social media discussions about the armed forces give an insight into the online community's attitude, which enables unfiltered, direct feedback for armed forces, something that did not exist before the social media era. Direct communication by members of the armed forces, in turn, can be used to transport personal views. However, this type of communication also carries the risk of conveying certain implicit ideas, clichés or role expectations, which can lead to stereotyping of the armed forces. In addition to these unconscious dangers inherent in the use of social media, social media can also be used in a targeted manner to harm and manipulate others, just as it has been done before with other media types (Porch 2002). Whether social media is an opportunity or a threat depends on the context and the way in which it is used, as well as on the information available in this context. In their contribution, *Social Media Use in Contemporary Armed Forces as a Mixed Blessing*, Jelena Juvan and Uroš Svete consider all these risks and chances, showing that social media is indeed a mixed blessing. They explain that there are opportunities as well as risks in social media both on an institutional and an individual level. On the individual level, social media facilitates communication with the home-front, which is an opportunity for soldiers. At the same time, it is also a danger because, depending on the (type and extent of) information they receive from home, soldiers can also become distracted from their mission, which could have fatal consequences. From an institutional point of view, on the other hand, social media is a valuable and cost-effective tool for recruitment. However, public sector organisations, such as armed forces, are not immune to being victims of hate speeches or shitstorms, which can affect their reputation and popular support.

Acknowledgements This anthology, and with it the international perspectives on this new field of research, could not have been realised without the excellent collaboration with the authors. It is their commitment to contribute with state-of-the-art research to the still new research area of social media and the armed forces that has enabled this book. Many of the authors were reached by means of the excellent network that exists through the Research Committee 01: Armed Forces and Conflict Resolution (RC01) of the International Sociological Association (ISA) and the European Research Group on Military and Society (ERGOMAS). We would like to express our sincere thanks for the exceptionally good and easy collaboration both with the research groups and the authors of this anthology. We also would like to thank Annelies Kersbergen of Springer for her dedicated collaboration.

We would like to thank the Military Academy (MILAC) at ETH Zurich for its constant commitment to promoting science, thereby making a valuable contribution to the exploration of new and innovative fields of research and the advancement of military sociology in Switzerland and internationally. Many thanks for the working time that MILAC has made available to the editors for the publication of this anthology.

Special thanks go to the reviewers who took the time and effort to read and review the various contributions. Their input is much appreciated by the authors and represents an added value for the development of the individual contributions and thus for this anthology. We would also like to thank Stefano De Rosa for being our sparring partner in discussing the book and its contributions. Finally, our gratitude goes to Anna Moser for the final revision of this book. She used her linguistic finesse to great effect, thus making the book even more worth reading.

References

Arkin W, Dobrofsky LR (1978) Military socialization and masculinity. J Soc Issues 34(1):151–168

Bandura A (1977) Social learning theory. Prentice Hall, Englewood Cliffs

Bandura A (2001) Social cognitive theory of mass communication. Mediapsychology 3:265–299

Bentele G (1994) Öffentliches Vertrauen – normative und soziale Grundlage für Public Relations. In: Armbrecht W, Zabel UJ (eds) Normative Aspekte der Public Relations. Grundlegende Fragen und Perspektiven. Eine Einführung. Springer, Wiesbaden, pp 131–158

Caddick N, Smith B, Phoenix C (2015) Male combat veterans' narratives of PTSD, masculinity, and health. Sociol Health Illn 37(1):97–111

DiMaggio PJ, Powell WW (1983) The iron cage revisited: institutional isomorphism and collective rationality in organizational fields. Am Sociol Rev 48(2):147–160

Fahmy S, Cho S, Wanta W, Song Y (2006) Visual agenda-setting after 9/11: individuals' emotions, image recall, and concern with terrorism. Vis Commun Q 13(1):4–15

Floridi L (2014) The 4th revolution. How the infosphere is reshaping human reality. Oxford University Press, Oxford

Fög – Forschungsinstitut Öffentlichkeit und Gesellschaft (2018) Qualität der Medien. Jahrbuch 2018. Hauptbefunde. Schwabe, Basel

Fög – Forschungsinstitut Öffentlichkeit und Gesellschaft (2019) Qualität der Medien. Jahrbuch 2019. Hauptbefunde. Schwabe, Basel

Hendricks VF, Vestergaard M (2019) Reality lost. Markets of attention, misinformation and manipulation. Springer, Cham

Hinton S, Hjorth L (2012) Unterstanding social media. SAGE, Los Angeles/London/New Delhi/Singapore/Washington, DC

Jacobs J (2016) Nutzung digitaler Medien und die Bundeswehr. In: Jacobs J, Zowislo-Grünewald N, Beitzinger F (eds) Social Media in der Lebenswelt und bei der Berufswahl Jugendlicher – Who cares? Nomos, pp 77–94

Katzenbach C (2017) Von kleinen Gesprächen zu grossen Öffentlichkeiten? Zur Dynamik und Theorie von Öffentlichkeiten in sozialen Medien. In: Klaus E, Drüeke R (eds) Öffentlichkeiten und gesellschaftliche Aushandlungsprozesse. Theoretische Perspektiven und empirische Befunde, Critical studies in media and communication, vol 14. transcipt, Bielefeld, pp 151–174

Maltby S, Thornham H (2016) The digital mundane: social media and the military. Media Cult Soc 38(8):1153–1168

Mühler K (2008) Sozialisation. Eine soziologische Einführung. Wilhelm Fink, Paderborn

Olsson E-K, Deverell E, Wagnsson C, Hellman M (2016) EU armed forces and social media: convergence or divergence? Def Stud 16(2):97–117. https://doi.org/10.1080/14702436.2016.1155412

Pavalanathan U, Datla V, Volkova S, Charles-Smith L, Pirrung M, Harrison J, Chappell A, Corley CD (2016) Discourse, health and well-being of military populations through the social media lens. In: AAAI Workshop: WWW and Population Health Intelligence

Porch D (2002) "No bad stories". The American media-military relationship. Nav War Coll Rev 55(1)

Reuters Institute (2019) Digital news report 2019. [Online.] University of Oxford, Oxford. [Viewed 20 February 2020]. Available from: www.digitalnewsreport.org

Rukavishnikov VO, Pugh M (2018) Civil-military relations. In: Caforio G, Nuciari M (eds) Handbook of the sociology of the military, Handbooks of sociology and social research, vol 19. Springer, Cham, pp 123–143

Schrape J-F (2011) Social Media, Massenmedien und gesellschaftliche Wirklichkeitskonstruktion. Berl J Soziol 21(3):407–429

Suchmann M (1995) Managing legitimacy strategic and institutional approaches. Acad Manag Rev 20(3):571–610

Wood JT (1994; 2000) Gendered lives: communication, gender, and culture, 4th edn. Wadsworth, Boston

Yılmaz M, Yılmaz U, Demir-Yılmaz EN (2019) The relation between social learning and visual culture. Int Electron J Elem Educ (IEJEE) 11(14):421–427

Part I
Social Media in the Everyday Life of Military Personnel

Ubiquity with a Dark Side: Civil-Military Gaps in Social Media Usage

Karin K. De Angelis, Ryan Kelty, Morten G. Ender, David E. Rohall, and Michael D. Matthews

Abstract Most college undergraduates are Millennials or Generation Z members. These generations are ferocious social media consumers across a range of platforms. Research exists on the U.S. military's adoption of social media, but less is known about the everyday implications of social media use and how service members might differ in their uses from their civilian peers. Using survey data comparing (N = 960) American civilian college students, Reserve Officer Training Corps cadets, and military academy cadets, we examine how these groups use social media, the educational and social impacts of this usage, and experiences with self-censorship and anxiety. We find a civilian-military gap when it comes to social media uses and experiences. Social media is more adverse for civilians than academy cadets in terms of time usage, impact on education, and experiences with the darker dimensions such as cyber-bullying and harassment. Civilians also practice less self-censorship of social media posts than cadets.

Keywords Social media · Time use · Academic impact · Cyberbullying · Civil-military gap

Readers should be aware that the views of the authors are their own and do not purport to represent the views of their respective academies, the Departments of the Air Force and Army, the Department of Defense, or the United States Government.

K. K. De Angelis (✉) · R. Kelty
United States Air Force Academy, Colorado Springs, CO, USA
e-mail: karin.deangelis@usafa.edu; ryan.kelty@usafa.edu

M. G. Ender · M. D. Matthews
United States Military Academy, West Point, NY, USA
e-mail: morten.ender@westpoint.edu; mike.matthews@westpoint.edu

D. E. Rohall
Missouri State University, Springfield, MO, USA
e-mail: drohall@missouristate.edu

E. Moehlecke de Baseggio et al. (eds.), *Social Media and the Armed Forces*, Advanced Sciences and Technologies for Security Applications,
https://doi.org/10.1007/978-3-030-47511-6_2

1 Introduction

The term "social media" originated in the 1990s (Bercovici 2010), and within a generation its use in North American culture had become ubiquitous (Danesi 2019). While to be human is to be social – we are unique in our abilities to read emotions, communicate with a shared language, and co-build shared cultures – the need to be face-to-face in social spaces no longer holds. Currently, we can express and experience our social nature through social media, which uses technologies for information sharing, and social networking sites, which allow for communication and interaction, typically within a bounded system of networks (boyd and Ellison 2007). We see the impact of social media at the micro and macro level from psychological, interpersonal, and institutional viewpoints as well as through beneficial connections (and challenges) across major social institutions, societies, and cultures.

In this chapter, we contribute to emerging research on social media use and its positive and negative consequences among university undergraduates in the United States. We focus on comparing civilian students with their Reserve Officer Training Corps (ROTC) and federal military academy peers. These three groups allow us to examine civilian-military differences across a range of contexts. Military academy cadets operate in a setting resembling aspects of a total institution (Goffman 1961) – meaning large groups of people working and residing together for considerable amounts of time in relative isolation from the larger community to which they belong. Cadets sleep, socialize, eat, attend classes, and complete physical training in the secure grounds of the academy and under its military regulations and ethos. ROTC cadets attend a civilian university, live in civilian housing and attend predominantly civilian classes, but also attend 'non-civilian' classes as well as military and physical training as part of their university's military detachment. Civilian students attend colleges and universities without directly engaging with military education or training, regardless of whether or not their school happens to have an ROTC detachment. Thus, military academy cadets and civilians represent the 'purer' ends of the continuum, while ROTC students exist at the intersection of the civilian and military environments. A total institution can be more or less usurped by social media, which might help us understand how undergraduates in these three contexts differ in their use and experience of social media.

In the following sections we review the literature on social media use and the military as well as trends in social media use among youth – including university students. We then present preliminary findings from our study examining civilian students, academy cadets, and ROTC cadets across a range of social media outcomes, including how they use social media, educational impacts, negative social impacts, and self-censorship and anxiety related to social media use. The chapter concludes with a discussion of the implications of our study's findings and thoughts on directions for future research.

1.1 The Military and Social Media

Although the United States military is at the forefront of many technological developments and adaptations, it can nevertheless lag behind mainstream society in adopting new technologies. However, as is the case for civilian society, once introduced, such technologies can quickly become ubiquitous, while the accompanying norms and values used to regulate technology-related behavior dawdle. Due to this cultural lag, the study of the military organization at the intersection of social media is profoundly new. Indeed, we challenge the reader to peruse, as did we, the index of books published before 2019 dealing with military topics and note that they mention 'social isolation' and 'social networking', but not 'social media'. Exceptions include Silvestri's (2015) book *Social Media in the American War Zone*, in which she conducts interviews and follows a small group of Marines and their Facebook posts. Her definition of social media, however, is narrow, examining only one specific medium – in this case Facebook – and how Marines communicate their deployment experiences. Similarly, retired General Martin Dempsey (with Brafman, Dempsey and Brafman 2018), the former Chief of the Joint Chiefs of Staff of the U.S. military, scratches the surface of social media with the idea of the 'digital echo' – providing practical examples of where social media has lasting and profoundly negative impacts because of its speed and ubiquity. This allows for potential misuse at the intersection of public and private spaces, potentially undermining group and leadership relations.

The U.S. military is a youth-oriented organization with approximately 66% of the active duty force being 30 years old or younger (Department of Defense 2017). This puts the majority of active duty personnel in the generational cohorts of Millennials and Generation Z,[1] with a much smaller proportion of Generation Xers (Dimock 2019) as well as a handful of senior Baby Boomer generals rounding out the organization. Young people – especially those from younger cohorts – are ferocious social media consumers (Dimock 2019). While all generations increasingly use social media, they differ on who uses which platforms (e.g., Facebook, Snapchat, etc.) and to what ends. For example, Generation Z (18–24 years old) uses Snapchat (78%) multiple times a day, while using Instagram (71%), Twitter (45%), and especially Facebook far less than Xers and Boomers (Smith and Anderson 2019). Moreover, Zs began using social media at a younger age than Millennials, thus making their use relatively life-long and ubiquitous. Such users are 'digital natives' who are fluent in the language and interactions connected to computers, the internet, and other screen technologies (Prensky 2001).

[1] Generations are typically a group of people born within the same time period. Scholars sometimes use the word "generation" to refer to a birth cohort. The following labels and birth years are typically associated generations in the United States: G.I. Generation (1901–1924); Silent Generation (1924–1945); Baby Boomers (1946–early 1960s); Generation X (early 1960s–1980); Millennials (1981–mid to late 1990s); and Generation Z (late 1990s–present) (See Ender et al. 2014).

Building on the intersection of technology and demographics, our research considers the extent to which people in the U.S. military use social media as well as the outcomes and experiences their social media use might produce. We compare university-affiliated adults of varying military attachment (i.e., ROTC cadets and service academy cadets in their early twenties due to age restrictions for officer commissioning) and non-military affiliated civilians (the vast majority in their early twenties)[2] in terms of preferred platforms, time usage, and consequences of use.

The research literature on U.S. military service members provides some limited studies on social media use in the military, with an overemphasis on military families rather than on the exclusive domain of the work life of service members, especially during deployments (Ender and Segal 1996, 1998; Ender 1995, 2005, 2009; Bell et al. 1999; Schumm et al. 2004). Research on service members' use of social media is even more sparse than research on the use by military families (Sherman et al. 2016; Semaan et al. 2017). Furthermore, a cultural lag has been identified as new social media are being adopted during war time, and ethics, policies, rules, and norms struggle to keep up with informal and formal regulations of behavior on these platforms (Wall 2010).

The U.S. experience should be understood within the broader context of western democracies' experience with the opportunities and challenges related to social media. The British have conducted studies on how their service members use social media in everyday ways and found that digital mundanity – that is, "…embodied, unthought and routine practices" – helps to inform and deflect military institutional identity (Maltby and Thornham 2016, 1165). The Swedes have examined blogs in a military-deployed context for use and hegemonic narrative content (Hellman and Wagnsson 2013; Hellman 2015) as well as similar use (Hellman et al. 2016) and patterns of content convergence (Olsson et al. 2016) among citizens of EU member states. The Canadians have adopted social media for institutional purposes, utilizing YouTube to promote the military organization (Mirrlees 2015). Their efforts focus on traditional recruitment for their all-volunteer military and on shaping the definition of the Canadian Armed Forces for both domestic and international audiences. Another study criticizes the political dimensions of social media use among military members (Lawson 2014). The authors claim that social media destabilizes dominant discourses about the military – in fact, social media may be *the* dominant discourse in an era of postmodern conflict, providing multiple voices and varied experiences and views related to the military (Lawson 2014). The Israelis are perhaps the furthest along in understanding the pervasiveness of social media in a forward deployed context and in communicating with citizens (Kuntsman and Stein 2015). Thus, the research literature is limited and spurious. However, it is clear that social media is a universal feature of military work life whose impact is yet to be fully examined and comprehended.

Another key transition brought about by the pervasiveness of social media is its accelerating tactical and strategic uses in military spaces (Gray and Gordo 2014).

[2] Our civilian samples are not restrictive – all eligible university students can complete the survey.

Online and networked military spaces are being expanded in new and novel ways (Maltby and Thornham 2016). Indeed, it is clear that current and future wars will be fought both on the ground and in cyberspaces (Gray 1997). Swedish blogging research about the mission in Afghanistan emphasized opportunities of social media use in a deployment context outweighing the risks (Hellman and Wagnsson 2013). However, we should be aware of the implications of sharing the horrors of war, out of context, with all constituents via social media (Tait 2008).

1.2 Social Media Trends Among Youth

As social media usage quickly became a common experience for youth, researchers scrambled to estimate the effects of these platforms on academic performance, social lives, mental health, and self-esteem. Early work focused on computer-based platforms such as MySpace, which had emerged as the most popular social networking platform by 2005, and instant messaging (boyd 2007). Spaces for online identity creation, performance, and sociality created time requirements of users as work was needed to build and maintain online personas, especially on social networking sites. Research suggests that social networking sites are becoming increasingly important in building and maintaining interpersonal relationships (Spies Shapiro and Margolin 2014). Users also have to consume the information available on social media (they are, after all, being voyeuristic in their consumption of information about others), which becomes especially important on sites that require interactions, reactions, and repeat performances. Thus, social media use can be time-consuming and represent lost opportunities with regard to other activities and interactions.

There is limited published empirical research available on the exact amount of time young adults spend on social media sites – although one implication is that screens, from laptops and tablets to smart phones, are now the main sites for media viewing. Tweens, teens, and young adults are more likely to use these devices to stream TV shows, movies, and music than they are stationary televisions or desktop computers, which makes their use mobile. They also use these screens to play video games. Media consumption often occurs alongside social media usage; in fact, such multi-tasking often is encouraged with chat rooms set up during video games and real-time responses to media events on social media sites. Taken together, researchers estimate that American teenagers spend around 9 h a day on media sites. For teens who are primarily social networkers, researchers estimate they spend slightly more than 3 h a day on social media (Common Sense Media 2015).

A number of scholars during the 1990s began examining the effects of the pervasiveness of communication media – including the dark side – on our lives (Gergen 1991; Meyrowitz 1993; Poster 1995; Turkle 1995). They chronicled the impact of communication media on our identity and selves, arguing that the western self would become mediated. Those works forecasted a coming social media pervasiveness including today's problems such as bullying, trolling, sexting, and catfishing.

Today's teens report that a majority of them (approximately 59%) have experienced some form of cyberbullying, most commonly name-calling and rumor-spreading (Anderson 2018). These experiences have consequences, especially on mental health.

These days, research, for instance by Twenge (2006, 2017), focuses on a generation that has come of age with social media. One key finding is the time and constancy which Generation Z spends on social media sites. There is pressure to participate and cultivate one's online persona in order to avoid experiencing FOMO (fear of missing out), which increases anxiety. Even though social media might portray happy, perfect lives, Twenge's (2017) research contributes to the growing literature highlighting a connection between social media use and depression among youth due to the constant demands and the frequent opportunities for social comparison (Fox and Moreland 2014; Lin et al. 2016).

Previous research indicates that military cadets differ little from their civilian peers across a range of attitudinal outcome measures – where sex and race provide greater explanatory power than institutional affiliation (Ender et al. 2014). However, we expect that the strict organizational structure and culture at the federal military academy may affect social media use and gratifications, causing these cadets to report different patterns of attitudes, behaviors and experiences compared to the social media experiences of ROTC cadets and civilian peers who attend civilian colleges and universities and are relatively more autonomous in their day-to-day activities. Given that these young collegians are the next generation of leaders in government, the economy, and civic life, we believe that these are some of the dominant areas in need of exploration at the intersection of the military and social media.

2 Method

The analysis is based on survey data collected from the Generation Z: Cadet and Civilian Attitudes study.[3] The overall study seeks to establish a database of attitudes in a specific segment of the U.S. population – military academy cadets, ROTC cadets, and civilian college students known as Generation Z.[4] In the broader study

[3] The previous study is known as the Bi-Annual Attitude Survey of Students (BASS). The GENZ is a continuation of the previous study with new and updated questions – specifically, questions related to social media not asked on the BASS (Ender et al. 2014).

[4] Military academies are all-inclusive and exclusive university level boarding schools where all cadets live, work, and study together. In the United States they include the United States Military Academy at West Point, NY, the United States Air Force Academy at Colorado Springs, CO, the United States Naval Academy in Annapolis, MD, and the United States Coast Guard Academy at New London, CT. The Reserve Officer Training Corps (ROTC) is another source of officer accessions. Officer Candidate School (OCS) and direct commissioning are two further sources. They consist of university or college student-cadets that attend civilian universities across the United States, participate in military education and training on campus, but live and study with their civilian peers.

we first examine attitudes toward a number of domains associated with military life, such as military professionalism, civil-military relations, the role of the military, the role of women in the military as well as social media questions and other social issues. Secondly, we compare all three groups regarding the most prevalent social problems facing their generation and the nation. Finally, we compare and contrast these attitudes among cadets at military academies, ROTC cadets, and civilian college students.

For this contribution, we examine and provide some preliminary analyses of our social media questions. This is an ongoing study that collects data from 100–200 individuals in each of the three target groups – military academy, ROTC, and civilian college students – each semester (twice per year). We utilize an availability sample. Cadets and civilians receive an email with an online link to the survey and participate on a voluntary basis. The survey is confidential, anonymous, and private.

Over a period of four academic semesters between fall 2017 and spring 2019, we surveyed about social media as part of a larger survey and have thus far obtained a subsample of students and cadets (N = 960) with 593 participants from a single military academy, 208 Reserve Officers' Training Corps (ROTC) cadets at multiple campuses, and 159 civilian college students at multiple campuses. The typical student in the military academy sample is male, white, nineteen or younger, leaning politically to the right, Christian, and in their first (freshman) year of college. ROTC cadets are close in proximity to the demographics of the military academy cadets. Civilian college students show far more diversity: a broader age range, a higher share of women, and more representative proportions regarding race and ethnicity.

2.1 Measures

Our analyses are based on numerous web-based survey questions about social media using close-ended Likert-type scales. We focus on three distinct areas. First, questions include items about time use, such as: how much time do you spend actively using social media? Is your level of social media use more or less than you intend? We ask questions about educational impact, for instance: does social media use during class distract you from class content? How often do you use social media while working on academic homework assignments? To what extent does homework take longer due to the intrusiveness of social media? Secondly, there are several questions that focus on the dark side of social media, asking whether or not one has been a target of cyberbullying, public shaming, and trolling during their undergraduate career. Finally, we ask a series of questions about censorship and anxiety related to social media use, such as: how often do you self-censor when posting your own content on social media? How often do you self-censor when responding to other people's posts on social media? How anxious are you about the possibility of being a future target of cyberbullying, public shaming, and/or trolling on social media?

2.2 Analysis Plan

Our analyses focus on bivariate frequency distributions of the key outcome variables outlined in the prior section by our three distinct groups – academy cadets (n = 593), ROTC cadets (n = 208), and civilian college students (n = 159).[5]

3 Results

Students across the three groups reported significant differences in the amount of time they spend on social media per day (one-way ANOVA, F = 57.769, p < .001). At the low end (no table), military academy cadets spend on average 1.87 h per day engaged with social media (s.d. = 1.68). ROTC cadets report spending 2.62 h on social media daily (s.d. = 2.57), more than three-quarters of an hour longer than their military academy peers. On the upper end of the distribution, civilian students report an average of 4.08 h per day on social media (s.d. = 3.40), which is more than double the average time reported by military academy cadets.

In addition to reporting the number of hours per day spent on social media, students and cadets also shared how this time related to their intended amount of time spent focused on social media (Table 1). Analysis confirms significant group differences between the intended and the actual time spent on social media (Chi-square = 26.913, $p < .05$). One in five civilian students reported spending significantly more time than they intended on social media. Military academy and ROTC cadets reported much lower levels of social media use that significantly exceeded their intentions, at 8.8% and 10.6% respectively. The trade-off that appears to push the distribution of civilian students' time on social media to exceed expectations out of

Table 1 Group comparison of the amount of time spent actively on social media relative to expectations[a]

	Academy	ROTC	Civilian
Significantly more time than I intend	8.8%	10.6	20.4
Much more time than I intend	19.4	15.7	21.7
Slightly more time than I intend	38.8	37.9	37.5
Just the right amount of time	23.2	25.3	16.4
Slightly less than I intend	2.2	2.0	0.7
Much less than I intend	1.4	1.0	0.0
Significantly less time than I intend	2.2	2.0	1.3
I do not use social media	4.0	5.6	2.0

[a]Chi-square = 26.913, $p < .05$

[5]The group Ns remain stable across each question with no more than 1–4 response differences based on missing values for each question.

line with their military academy and ROTC peers is at a level with "just the right amount of time". Approximately one quarter of military academy and ROTC cadets reported spending just the right amount of time on social media, while only 16.4% of civilian students indicate this level of use. Comparisons of social media use relative to expectations are much more harmonious across the three groups at the other use levels. Very small percentages of each group use social media less than they intend to or abstain from social media completely. It is not to be missed that less than 5% of each group reported not using social media at all – thus, social media is hyper-pervasive among our samples.

Cadets' and students' experiences with social media and its impact on academic performance are reported in Table 2. Roughly a third of academy cadets reported never being distracted in class by social media. The proportion of ROTC cadets reporting never being distracted is considerably lower (23.3%), with civilians being lower still (19.7%). The pattern of increased distraction indicates that academy and ROTC cadets range from a quarter to a third of respondents across the three least frequent disruption categories, whereas civilian students fall below 20% in the "never" and "rarely" categories and then spike at 36% in the "sometimes" category. Civilians also reported much higher proportions of usage categorized as "often" and "very often", adding up to a total of a quarter of respondents. By comparison, just over 10% of academy cadets reported these highest levels of academic distraction, with ROTC cadets splitting the difference between groups at approximately 17%. These results represent significant differences between the civilian and military groups (Chi-square = 33.694, $p < .01$).

When asked to what degree it takes longer to complete homework due to the intrusiveness of social media, greater parity across groups is observed, although results continue to indicate significant group differences (Table 2;

Table 2 Educational impacts of social media use among undergraduates

		Academy	ROTC	Civilian
In the last year, used social media during class and it distracted me from class content.[a]	Never	33.6%	23.3	19.7
	Rarely	25.2	25.1	19.1
	Sometimes	29.1	35.0	36.3
	Often	9.9	11.8	16.6
	Very often	2.2	4.9	8.3
Doing homework takes longer due to the intrusiveness of social media.[b]	Strongly disagree	2.6	5.1	0.7
	Disagree	5.8	8.1	10.6
	Slightly disagree	5.3	9.1	7.9
	Neither agree nor disagree	11.5	12.2	11.3
	Slightly agree	28.8	25.9	19.2
	Agree	25.4	19.3	26.5
	Strongly agree	20.6	203.	23.8

[a]Chi-square = 33.694, $p < .01$
[b]Chi-square = 21.843, $p < .05$

Chi-square = 21.843, $p < .05$). More than 20% of respondents in each group strongly agreed with this statement.

Approximately one fifth to one quarter of respondents in each group also indicated agreement or slight agreement with this statement, with academy cadets showing slightly higher proportions than their ROTC and civilian peers. On the other end of the distribution, 2.6% of academy cadets and 5.1% of ROTC cadets reported that social media never distracts them from their homework, whereas less than 1% of civilian students reported the absence of social media distraction when completing their homework. Taken together, more than two thirds of respondents across all groups indicated some level of social media interference with their homework. The pattern shows that civilian students find social media the most distracting in completing their academic work, followed by ROTC cadets, and then academy cadets.

Experiences with being a target of public shaming, cyberbullying, and/or trolling as an undergraduate are summarized in Table 3. Results show that while these experiences are relatively uncommon, between 10 and 20% of respondents from each group report positive or probable (i.e., maybe) experiences with these negative outcomes of social media. Among civilian students, 11.2% indicate they have been publicly shamed, while 8.6% report being cyberbullied as undergraduates. These values are more than twice the proportion of academy and ROTC cadets who reported being the target of these social media behaviors (Shaming Chi-square = 13.459, $p < .01$; Cyberbullying Chi-square = 20.017, $p < .01$). The results for trolling suggest that while differences are observed, the three groups do not differ significantly from one another (Chi-square = 6.213, NS). ROTC cadets and civilian students are more similar, with 8.1% and 7.9% indicating having been a victim of this behavior, respectively. 4.5% of academy cadets reported experiencing trolling, only half of their ROTC and civilian peers. In general, civilian students appear to be most likely, whereas academy cadets appear to be least likely to experience these negative effects of social media.

Table 3 Student experiences with the dark side of social media: public shaming, cyberbullying, and trolling

		Academy	ROTC	Civilian
I have been a victim of public shaming via social media as an undergraduate.[a]	Yes	4.3%	3.5	11.2
	No	88.9	89.4	80.9
	Maybe	6.9	7.1	7.9
I have been a victim of cyberbullying as an undergraduate.[b]	Yes	2.9	3.0	8.6
	No	90.5	94.9	82.2
	Maybe	6.5	2.0	9.2
I have been a victim of trolling as an undergraduate.[c]	Yes	4.5	8.1	7.9
	No	89.2	86.9	84.2
	Maybe	6.3	5.1	7.9

[a]Chi-square = 13.459, $p < .01$
[b]Chi-square = 20.017, $p < .01$
[c]Chi-square = 6.213, NS

Our final set of analyses focused on three censorship and anxiety items related to social media (Table 4). The minor differences observed across the groups for these outcome variables do not reach significance. More academy (40.9%) and ROTC (42.3%) cadets than civilian students (33.3%) indicated that they "always" censor themselves when posting to social media. Conversely, civilian students (20.6%) are more likely to self-censor their own posts only "sometimes" compared to both academy (15.5%) and ROTC (14.8%) cadets. Nearly 10% of both civilian students and ROTC cadets reported "never" censoring themselves when posting on social media, which is nearly double the rate of academy cadets (5.3%).

When asked how often one self-censors when responding to another's post on social media, we see essentially the same pattern of results presented above for self-censoring. Academy and ROTC cadets reported higher rates of "always" engaging in self-censorship when responding to others (39.7% and 43.3%, respectively), compared to civilians who trail behind at 33.1% for this level of self-censorship. ROTC cadets (10.5%) had the highest proportion of individuals who reported that they "never" self-censor when responding to others' social media posts, followed by 7.3% of academy cadets and just 6.3% of civilian students. These results indicate that ROTC cadets engage in similar levels of consistently refusing to self-censor in

Table 4 Group comparisons on censorship and anxiety related to social media among undergraduates

		Academy	ROTC	Civilian
How much do you censor yourself when posting on social media due to concerns about how others may respond to you or view you?[a]	Always	40.9%	42.3	33.3
	Most of the time	30.8	26.9	27.0
	About half the time	7.6	6.0	9.9
	Sometimes	15.5	14.8	20.6
	Never	5.3	9.9	9.2
How much do you censor yourself when responding to others on social media due to concerns about how others may respond to you or view you?[b]	Always	39.7	43.3	33.1
	Most of the time	30.2	25.4	30.3
	About half the time	8.4	6.6	9.9
	Sometimes	14.5	14.4	20.4
	Never	7.3	10.5	6.3
To what degree are you anxious about being a victim of trolling, bullying, or public shaming in the future due to having a social media presence?[c]	Always	5.4	4.3	8.0
	Most of the time	5.2	3.8	5.1
	About half the time	6.1	4.3	10.1
	Sometimes	18.6	22.6	26.1
	Never	64.7	65.1	50.7

[a]Chi-square = 11.908, NS
[b]Chi-square = 9.204, NS
[c]Chi-square = 12.505, NS

their own posts and when responding to others. Academy cadets are slightly more likely to "never" self-censor when responding to posts than in their own posts, whereas the converse is true for civilian students.

Responses to the question of how anxious cadets and students where about being a future victim of trolling, cyberbullying or public shaming on social media indicate generally low levels of anxiety across all groups (Table 4). Seventy five to almost 90% of respondents across the three groups indicated either "never" or only "sometimes" feeling anxious. Both academy and ROTC cadets reported essentially similar levels of anxiety about being targeted on social media, with two-thirds indicating "never" being anxious, and approximately one fifth indicating "sometimes" being anxious. Conversely, only half of the civilian students reported "never" being anxious about being targeted on social media, with another quarter indicating "sometimes" being anxious. Regarding anxiety levels of "about half the time", "most of the time", and "always", academy and ROTC cadets reported approximately 4–6% at each level, whereas civilians reported between 5 and 10% at each level.

Taken together, these results show a general pattern that Academy and ROTC cadets reported engaging in higher levels of self-censorship on social media than civilian students, with all groups indicating high levels of self-censorship. Additionally, civilian students reported somewhat greater anxiety over being trolled, bullied or shamed on social media – though all groups have a strong majority who reported no to minimum anxiety.

4 Implications and Conclusion

In this paper we contribute to an emerging literature on the uses and positive and negative consequences of social media use among university undergraduates in the United States with a specific focus on comparing civilian students with their ROTC and federal military academy peers. These three groups allow us to examine civilian-military differences across a range of (non)military contexts – moving across relative total institutional contexts.

The study presented here provides some preliminary analyses of the univariate results. Overall, there appears to be a civilian-military cleft when it comes to social media uses and gratifications among (non)military-affiliated U.S. undergraduates, with social media being more detrimental to civilians than cadets at the military academy. Yet virtually all the undergraduates are impacted by social media, and all groups are like-minded about the future of social media.

A more nuanced observation shows seven major findings from our study. First, the more civilian, the more daily hours are spent on social media. Second, the more civilian, the more time is spent on social media than actually intended/desired. Third, the more civilian, the higher are the in-class distractions from social media. Fourth, the more civilian, the more distracted students are by social media when conducting homework. Fifth, the more civilian, the more students have had experiences with the dark side of social media such as public shaming, cyberbullying, and trolling. Sixth, the more civilian, the less self-censoring occurs regarding posts on social media. Last, all three groups share *similar* feelings about being future victims of the dark side of social media, with civilians showing slightly greater anxiety. Thus, the preliminary results show a civil-military difference: the more civilian, the more pervasive social media is in the lives of undergraduate students. Stated another way, the more military an undergraduate, the less pervasive social media is in their everyday life.

Thus, the sociological question changes as follows: what is it about military undergraduate life that mediates social media use? To begin with, social structure may play a role. The lifeworld of military academy cadets and – somewhat less but certainly more than that of civilians – ROTC cadets is more structurally demanding both day-to-day and across the entire undergraduate experience. Military academy institutions remain more total compared to civilian universities (Goffman 1961). For example, all cadets have more strenuous academic, physical, and military requirements than most civilian students (which is not to say that there are no undergraduates with highly demanding day-to-day schedules, such as full and part-time work, family commitments, athletic requirements, and rigorous courses either self- and/or externally imposed). All cadets are required to be fulltime students with full class attendance and less curricular choices – oriented toward graduating in eight semesters/47 months – and all have athletic and military responsibilities and requirements. Additionally, cadets have socially restricted lives as dictated by the total institution, such as limited campus privileges and daily 11 pm lights out in the dormitories where all cadets reside. Thus, with time being a zero-sum entity – there are only so many hours in a day – cadets simply cannot find the same downtime compared to their civilian peers who have relatively more autonomy to spend time on social media after they have completed the demands of the institution.

Another explanation might be subcultural. Cadets who self-select into the military may possess a different cultural disposition than their civilian peers – they might be less oriented toward social media and its popular civilian cultural trends than civilians. In this way, cadets represent an ethos and subculture of American society that has less interest in the musings found on social media.

Future research should address these topics with greater multivariate focus. First, questions might include whether there are subgroups that vary in how they are impacted by social media use. Previous research has found that sex and politics are better predictors of attitudes than degrees of military affiliation among these groups (Ender et al. 2014, 2016; Laurence et al. 2017; Matthews et al. 2009; Rohall et al. 2006). Might this hold true for social media where military affiliation becomes increasingly less significant as more important variables are controlled? Ultimately, are there pockets of subgroups that are social media misusers?

Secondly, there is the question of how social media relates to personal gratifications and mental health outcomes. Research suggests that screen time negatively correlates with mental health. For example, Twenge (2017) reports that the higher the amount of screen time, the higher the number of young people who report mental health issues such as depression, loneliness, and suicidal ideation among other negative mental health outcomes. In our study, we have measures for self-esteem, life satisfaction, and general well-being.

Finally, our data has multiple limitations. For one thing, the data is limited to convenience/availability samples for all three groups, meaning that generalizability is limited here. For another thing, our data represents but one moment in time. We need to continue monitoring these social media uses where abatement of social media is not the norm (Smith and Anderson 2019), keeping in mind that a disproportionate number of both academy and ROTC cadets come from the civilian sector. As younger generations become more comfortable with handheld technology and the technology itself becomes more efficient (e.g., Apple watches), will their use and gratifications follow them into university? Moreover, the data is based on self-report social media usage and experiences. All three undergraduate groups could be under-reporting a perceived social problem/addiction such as social media usage and/or might be in denial about cyberbullying or harassment issues. Indeed, there may even be nuanced differences among the three groups in their self-reporting. Yet, the results do offer some preliminary empirical insights into the three groups and their social media usage.

In terms of implications, we can offer two suggestions at this preliminary stage. First, what can civilian institutions learn from military academies to structurally reduce preoccupations with social media among civilian college students? Second, what types of initiatives might be implemented that can educate all students and cadets to self-regulate themselves? For example, data from the present study could be with students to inform them about their usage in order to help them with their personal time-management skills. In this regard, Google, for instance, offers a technological fix for social media use: the Google Analytics app for smart phones offers daily and weekly analytics of cellphone usage and behavior including bounce rates, pageviews, and pages and sessions.

Social media is pervasive among our sample groups – less than 5% of each group report *not* using social media. Social media is here to stay, at least for the immediate future (Smith and Anderson 2019). Its ubiquity intrudes on the lives of students and cadets and impacts their ability to focus and concentrate as well as delivering and creating anxiety – anxiety students do not need to compound the already profound stress imposed by university study. Furthermore, excessive social media use may not be setting a productive precedent for students as working citizens and future leaders both in military and civilian society.

References

Anderson M (2018) A majority of teens have experienced some form of cyberbullying. Pew Research Center [online]. Available from: https://www.pewinternet.org/2018/09/27/a-majority-of-teens-have-experienced-some-form-of-cyberbullying/

Bell DB, Schumm WR, Scott B, Ender MG (1999) The desert FAX: calling home from Somalia. Armed Forces Soc 25(3):509–521

Bercovici J (2010) Who coined 'social media'? Web pioneers compete for credit. Forbes (December 9) [online]. Available from: https://www.forbes.com/sites/jeffbercovici/2010/12/09/who-coined-social-media-web-pioneers-compete-for-credit/#1b3e9dc51d52
boyd D (2007) Why youth (heart) social network sites: the role of networked publics in teenage social life. In: Buckingham D (ed) MacArthur Foundation series on digital learning – youth, identity, and digital media volume. MIT Press, Cambridge, MA
boyd D, Ellison NB (2007) Social network sites: definition, history, and scholarship. J Comput-Mediat Commun 13(1):210–230
Common Sense Media (2015) The common sense census: media use by tweens and teens [online]. [Viewed June 25, 2019]. Available from: https://www.commonsensemedia.org/research/the-common-sense-census-media-use-by-tweens-and-teens
Danesi M (2019) Popular culture: introductory perspectives, 4th edn. Rowman and Littlefield, Lanham
Dempsey M, Brafman O (2018) Radical inclusion: what the post-9/11 world should have taught us about leadership. Missionday
Department of Defense (DoD), Office of the Deputy Assistant Secretary of Defense for Military Community and Family Policy (2017) 2017 demographics: profile of the military community
Dimock M (2019) Defining generations: where millennials end and generation Z begins [online]. Available from: https://www.pewresearch.org/fact-tank/2019/01/17/where-millennials-end-and-generation-z-begins/
Ender MG (1995) G.I. phone home: the use of telecommunications by the soldiers of operation just cause. Armed Forces Soc 21(3):335–334
Ender MG (2005) Divergences in traditional and new communication media use among army families. In: Ouellet E (ed) New directions in military sociology. de Sitter Publications, Whitby, pp 255–295
Ender MG (2009) American soldiers in Iraq: McSoldiers or innovative professionals? Routledge, New York/London
Ender MG, Segal DR (1996) V(E)-mail to the foxhole: isolation, (tele)communication, and forward deployed soldiers. J Polit Mi Sociol 24(1):83–104
Ender MG, Segal DR (1998) Cyber-soldiering: race, class, gender and new media use in the military. In: Ebo B (ed) Cyberghetto or cybertopia: race, class, and gender on the internet. Praeger Publishers, Westport, pp 65–82
Ender MG, Rohall DE, Matthews MD (2014) The millennial generation and national defense: attitudes of future military and civilian leaders. Palgrave Pivot, Basingstoke/New York
Ender MG, Rohall DE, Matthews MD (2016) Cadet and civilian undergraduate attitudes toward transgender people: a research note. Armed Forces Soc 42(2):427–435
Fox J, Moreland JJ (2014) The dark side of social networking sites: an exploration of the relational and psychological stressors associated with Facebook use and affordances. Comput Hum Behav 45:168–176
Gergen KJ (1991) The saturated self. Basic Books, New York
Goffman E (1961) Asylums: essays on the social situation of mental patients and other inmates. Anchor Books, New York
Gray CH (1997) Postmodern war: the new politics of conflict. Guilford Press, New York
Gray CH, Gordo ÁJ (2014) Social media in conflict comparing military and social-movement technocultures. Cult Polit 10(3):251–261
Hellman M (2015) Milblogs and soldier representations of the Afghanistan War: the case of Sweden. Media, War Conflict 9(1):43–57
Hellman M, Wagnsson C (2013) New media and the war in Afghanistan: the significance of blogging for the Swedish strategic narrative. New Media Soc 17(1):6–23
Hellman M, Olsson E-K, Charlotte Wagnsson C (2016) EU armed forces' use of social media in areas of deployment. Media Commun 4(1):51–62
Kuntsman A, Stein RL (2015) Digital militarism: Israel's occupation in the social media age. Stanford University Press, Stanford

Laurence JH, Milavec BL, Rohall DE, Ender MG, Matthews MD (2017) Predictors of support for women in military roles: military status, gender, and political ideology. Milit Psychol 28(6):488–497 [online]. Available from: http://www.apa.org/pubs/journals/mil/index.aspx

Lawson S (2014) The US military's social media civil war: technology as antagonism in discourses of information-age conflict. Camb Rev Int Aff 27(2):226–245

Lin LY, Sidani JE, Shensa A, Radovic A, Miller E, Colditz JB, Hoffman BL, Giles LM, Primack BA (2016) Association between social media use and depression among U.S. young adults. Depress Anxiety 33:323–331

Maltby S, Thornham H (2016) The digital mundane: social media and the military. Media, Cult Soc 38(8):1153–1168

Matthews MD, Ender MG, Laurence J, Rohall DE (2009) Role of group affiliation and gender attitudes toward women in the military. Mil Psychol 21(2):241–251

Meyrowitz J (1993) No sense of pace: the impact of electronic media on social behavior. Oxford University Press, New York

Mirrlees T (2015) The Canadian Armed Forces "YouTube war": a cross-border military-social media complex. Glob Media J 8(1):71–93

Olsson E-K, Deverell E, Wagnsson C, Hellman M (2016) EU armed forces and social media: convergence or divergence? Def Stud 16(2):97–117

Poster M (1995) The second media age. Polity Press, Cambridge

Prensky M (2001) Digital natives, digital immigrants Part 1. On the Horizon 9(5):1–6

Rohall DE, Ender MG, Matthews MD (2006) The effects of military affiliation, gender, and political ideology on attitudes toward the wars in Afghanistan and Iraq. Armed Forces Soc 33(1):59–77

Schumm WR, Bell DB, Ender MG, Rice RE (2004) Expectations, use, and evaluations of communications media among deployed peacekeepers. Armed Forces Soc 30(4):649–662

Semaan B, Britton LM, Dosono B (2017) Military masculinity and the travails of transitioning: disclosure in social media. In: Proceedings of the 2017 Conference on Computer Supported Cooperative Work and Social Computing (Feb/Mar), pp 387–403 [online]. Available from: https://dl.acm.org/citation.cfm?id=2998221

Sherman MD, Rudi JH, Westerhof L, Borden LM (2016) Social media communication among military spouses: review of research and recommendations for moving forward. Mil Behav Health 4(4):325–333

Silvestri LE (2015) Friended at the front: social media in the American war zone. University Kansas Press, Lawrence

Smith A, Anderson M (2019) Social media use in 2018. Pew Research Center (March 1) [online]. Available from https://www.pewinternet.org/2018/03/01/social-media-use-in-2018/

Spies Shapiro LA, Margolin G (2014) Growing up wired: social networking sties and adolescent psychosocial development. Clin Child Fam Psychol Rev 17(1):1–18

Tait S (2008) Pornographies of violence? Internet spectatorship on body horror. Crit Stud Media Commun 25(1):91–111

Turkle S (1995) Life on the screen: identity in the age of the internet. Simon & Schuster, New York

Twenge J (2006) Generation me: why today's young Americans are more confident, assertive, entitled – and more miserable than ever before. Free Press, Chicago

Twenge J (2017) iGen: why today's super-connected kids are growing up less rebellious, more tolerant, less happy – and completely unprepared for adulthood – and what that means for the rest of us. Atria Books, New York, London, Toronto, Sydney, New Delhi

Wall M (2010) In the battle(field): the US military, blogging and the struggle for authority. Media, Cult Soc 32(5):863–872

Social Media Use at a U.S. Military Academy: Perceived Implications for Performance and Behavior

Sara Beth Elson, Ryan Kelty, Keith Paulson, John Bornmann, and Karin K. De Angelis

Abstract A growing body of research highlights the impacts of social media use for civilians. However, few studies have examined the possible implications of social media use for military personnel. In order to fill this gap, we surveyed cadets and faculty at a United States service academy. Most cadets and faculty members agreed that cadets spend more time than is ideal on social media, and most faculty surveyed had observed cadets using social (and other) media during lectures. Both cadets and faculty agreed that social media use impedes academic performance. However, social media use also had positive implications, as cadets listed connecting with friends and family as motives for using these media. Separately, cadets ranked the social media app Jodel among their top five most popular – in contrast to civilians. Jodel affords anonymity and, as such, potentially enables cadets to post controversial messages without fear of identification and accountability. Possible implications of these findings for the academy and the armed forces are discussed.

Keywords Social media · Social media and military · Social media military performance · Social media military readiness · Social media behavioral impact · Jodel

---

We gratefully acknowledge the contributions of Hal Greenwald, who is now at the Air Force Office of Scientific Research. Hal contributed while a MITRE employee. We also gratefully acknowledge the contributions of John Kilgo and Maura Tennor, both of whom work at MITRE. Approved for Public Release; Distribution Unlimited. Public Release Case Number 20-0028 PA#: USAFA-DF-2020-8 The views expressed in this contribution are those of the authors and do not necessarily reflect those of the U.S. Air Force Academy, U.S. Air Force, or the Department of Defense.

S. B. Elson (✉) · K. Paulson · J. Bornmann
The MITRE Corporation, Bedford, MA, USA
e-mail: selson@mitre.org

R. Kelty · K. K. De Angelis
United States Air Force Academy, Colorado Springs, CO, USA

E. Moehlecke de Baseggio et al. (eds.), *Social Media and the Armed Forces*, Advanced Sciences and Technologies for Security Applications,
https://doi.org/10.1007/978-3-030-47511-6_3

1 Introduction

At present, the youngest generation of personnel entering the armed forces and service academies is digitally native – never having known life without the internet. The Pew Research Center has dubbed the current college-age population as "Generation Next" – a generation that is more digitally active than any previous generation (Kohut et al. 2007; Madden and Jones 2002; Salaway et al. 2008). Although a growing body of research highlights the implications of social media use among young adults, especially for mental health and academic performance, no previous research (to our knowledge) has examined the possible implications of social media use for professional readiness at military service academies. As a first step to fill this gap, we examined social media use at a United States service academy as well as the potential implications of this use for professional productivity and readiness. We note that the findings of this study apply to the American context and not necessarily to that of other countries.

The most closely-related research context for the current work consists of research on social media use at civilian universities and colleges – and the implications of that use for academic performance. We will first examine what is known, to this point, from the research conducted at civilian universities. Next, we will address the question of whether the effects of social media use at a service academy resemble those at civilian universities. Finally, we will discuss the potential implications of social media use for the unique circumstances of the military.

Regarding students at civilian universities (and teenagers in the civilian population, more broadly), although research has yielded contradictory findings, certain patterns are emerging – notably the increasing use of social media. For example, the UCLA Higher Education Research Institute has conducted the Freshman Survey for nearly three decades, asking incoming freshmen to report how many hours per week they spend doing a variety of activities. From 2007 to 2015, around a quarter of freshmen consistently reported spending six or more hours per week on social media. However, in 2016 the proportion of freshmen reporting social media use for at least 6 h per week jumped to 40.9% - nearly 14 percentage points higher than the previous high of 27.2% reached in both 2011 and 2014 (Eagan et al. 2016). In another recent survey focusing on teens, the Pew Research Center found that 45% say they are online almost constantly (Anderson and Jiang 2018) – suggesting the possibility that this includes during class.

Focusing on the implications of social media use at civilian universities, Doleck and Lajoie (2018) examined 23 peer reviewed articles from 2008 through 2016 and found that roughly two thirds of the articles in their data set reported that social media use predicts poorer academic performance. However, Doleck and Lajoie caveat their findings due to a lack of clear definitions of "social media," which sites were used for the study, and which types of behaviors were examined. In the present article, we distinguish between social media use for non-academic purposes versus social media used as an integral part of academic tasks, focusing specifically on the former.

Notwithstanding the caveats to Doleck and Lajoie's findings, a similar meta-analysis (Liu et al. 2017) of 28 studies focusing on adolescents and young adults found that increased social media use predicted poorer academic performance (i.e., grade point average or GPA) across most of the studies included. According to Liu et al., one way in which social media use has been found to impair academic performance lies in the multitasking involved with using social media while studying. Multitasking is the simultaneous execution of two or more information-processing activities (i.e., cognitive processes) at the same time. According to several studies, multitasking by using a social media site while studying distracts attention from learning, thereby impairing students' overall semester GPAs (Karpinski et al. 2013; Kirschner and Karpinski 2010; Judd 2014; Junco 2012; Junco and Cotten 2012; Golub and Miloloža 2010). For example, Golub and Miloloža (2010) examined multitasking and the intensity of Facebook use (i.e., a combination of variables accounting for frequency of logins, duration of sessions, number of friends, attitudes toward the use, and others). In a hierarchical regression analysis, Golub and Miloloza first found that the intensity of Facebook use made a significant contribution to the prediction of a negative Facebook impact on academic performance. In a later step, they found that active multitasking made a significant contribution above and beyond the contribution of the intensity of Facebook use.

Increased reliance on social media sites has also been found to predict reduced sleep quality and increased everyday cognitive failures (Orzech et al. 2016; Xanidis and Brignell 2016). In particular, students who reported spending more time on social media sites before bedtime were found to have more difficulty with their school assignments than those who did not, because they were always cognitively fatigued when in class or trying to study. Excessive social media use may have also taken time away from the actual task of studying.

The effects of multitasking on performance might be particularly pronounced among younger college students, as found by Junco (2015). Specifically, among first-year college students, the amount of time spent on Facebook predicted lower GPAs in a way that it did not for students in higher grades. When it came to multitasking with Facebook (as opposed to the amount of time spent using it), multitasking predicted lower GPAs for first-years, sophomores, and juniors, but not for seniors. Seniors spent significantly less time on Facebook and less time multitasking with Facebook than did students in the lower grades.

It should be noted, however, that the relationship between social media (and, in particular, Facebook) use and academic performance has been found to depend on the type of test used to measure academic performance. For example, when researchers examined literacy tests, they found that greater Facebook use predicted better performance (Alloway et al. 2013). Alloway et al. (2013), for instance, found that students who had been using Facebook for over one year achieved significantly higher scores in working memory, verbal ability, and spelling compared with those who had been using Facebook for less than a year. In that sense, Facebook use functioned as a working memory task that trained students' information-processing ability. The researchers speculated that the prolonged use of social media sites like

Facebook might give students training that can boost cognitive skills and affect literacy test performance, including verbal and spelling scores.

1.1 Use of Media/Internet During Class

Focusing on media/internet use in general, several researchers have examined the effects of such use during class. For example, Ravizza et al. (2017) assessed the relationship between laptop internet use (both for academic and non-academic purposes) and classroom performance. In addition, they assessed whether intelligence, motivation, and interest in the course material could account for any relationship between internet use and performance. Their results showed that non-academic internet use was common among students who brought laptops to class and that higher levels of non-academic internet use predicted poorer class performance. This correlation could not be accounted for by motivation, interest, or intelligence. In addition, class-related internet use conferred no benefit on classroom performance.

As one might guess, similar results emerged from research on cell phone use during class. Bjornsen and Archer (2015) examined the correlation between daily in-class cell phone use and test grades among college students. They found that greater cell phone use predicted lower test scores, regardless of students' sex and GPAs. Concurring with all of the findings discussed above, Ellis et al. (2010) found that multitasking during class caused lower grade performance among undergraduate business students. Given the above findings regarding internet use during class, we decided to examine any reported effects that classroom media use might have on academic performance at a military academy.

In the following, we will examine the structure of military academies versus civilian universities and the possible implications of social media use at military academies.

1.2 Structure of Military Academies Versus Civilian Universities in the United States

In the United States, military academies have a significantly different structure compared to civilian universities. Some of the crucial differences are centered around four major areas: military command structure, non-academic education/training, time constraints for graduation, and living within a closed system. Regarding command structure, military academies sort cadets into multiple military units, e.g. squadrons, companies, and others. These military units generally rank cadets in upper grades higher than those in lower grades, assigning specific command roles to cadets. The units are normally supervised by actual military officers and trainers, who serve as mentors and role models to the cadets.

In addition to the educational curriculum, academies include a significant non-academic educational component tied to military and physical training. These courses, events, and even participation in sports (intercollegiate and clubs) are considered an essential part of the Academy experience. Moreover, the non-academic components are required for graduation and commissioning officers.

Military academies, due to law, regulation, and cost concerns, generally (with very few exceptions) require cadets to complete their chosen degrees and non-academic requirements in 4 years. Cadets receive grades and scores tied to their performance across academic, military, and physical components. The academies do not allow cadets to enter at any level other than freshman (i.e., cadets with previous higher education credits may transfer credits, but will still enter as freshmen and be required to complete 4 years at the academy). The academies are also closed systems as opposed to traditional, more open-system civilian universities. Cadets live exclusively on campus and need a pass or permission to leave campus. This practice has traditionally resulted in a more tightly-knit community, as almost all interaction occurs within the boundaries of the military academy's campus.

Although several factors differentiate military academies and civilian universities, they do share certain qualities. Namely, both provide academic educations and degrees of the students' choice. Both may include an honor code, although studies at civilian universities did not consistently find that honor codes decrease cheating (Konheim-Kalkstein et al. 2008). Within military academies, honor codes are a core part of cadets' obligations and are enforced more strictly than at civilian universities – a single infraction of an honor code can result in disenrollment. The honor code at a military academy also requires adherence to a military chain of command, paralleling the broader military context in which academies reside. To this point, it is not known which, if any, implications social media use might have for academic performance at service academies. The present paper takes a step toward addressing this question.

1.3 Social Media Use and Academic Performance at Military Service Academies

Although a growing body of research has examined the implications of social media use for academic performance at civilian universities (e.g., the studies discussed above, such as Eagan et al. 2016; Doleck and Lajoie 2018; Liu et al. 2017; Karpinski et al. 2013; Kirschner and Karpinski 2010; Judd 2014; Junco 2012; Junco and Cotten 2012; Golub and Miloloža 2010 and others), no such studies – to our knowledge – have been conducted at military service academies. In the United States, service academies provide cadets not only with an intensive academic curriculum, but also with a demanding program of military leadership preparation as well as athletic training. Cadets are preparing to become military leaders in highly regimented and isolated environments. Like their civilian counterparts, cadets might turn toward social media as a means of connecting with friends, family, and other cadets. However, at this point the extent and nature of social media use among cadets are not yet known. Nevertheless, regardless of the similarities and differences between cadets and students at civilian universities, it can be assumed that any negative effects of social media use on academic performance at service academies have potential implications for the professional readiness of future military leaders.

Previous research on social media use in military settings has examined such use by service members during deployment (e.g., Skopp et al. 2016). Results of this research show that Facebook offers deployed service members an opportunity to connect regularly with family members and friends, which may help bolster perceived social support and deployment resilience. Indeed, several previous studies in the civilian population have found that using social media can play a helpful role in building social capital – something that can be potentially helpful in the military and at service academies (e.g., Liu et al. 2016; Liu and Brown 2014; Ellison et al. 2007). However, Skopp et al. (2016) note that the daily connection to home life and its problems during deployment or the problematic use of Facebook may result in a loss of camaraderie and other adverse outcomes. Furthermore, when deployed, service members are on structured schedules in a dangerous environment and only have access to a limited range of activities. Their attention to the mission is critical to force readiness. Although cadets at a service academy are not deployed, their lives may resemble those of deployed personnel in certain respects, such as time pressures, structured schedules, attention to a mission, access to a limited range of activities, and, importantly, isolation from family and friends outside the academy. The use of social media and, in particular, Facebook might play a role in promoting cadets' mental health and resilience, but a question arises regarding the point at which such use starts to interfere with academic achievement.

1.4 Behavioral Implications of Social Media Use at Service Academies

In addition to the potential effects on academic achievement, the use of social media could potentially have a variety of behavioral implications – both positive and negative – which have not been examined at service academies. On the one hand, as noted above, previous studies have found that using social media can play a helpful role in building social capital. On the other hand, social media has also been used for cyber-bullying (Dooley et al. 2009; Harrison 2018).

According to Dooley et al. (2009), cyber-bullying comprises a set of aggressive behaviors that are enacted via electronic media. A study by Vandebosch and van Cleemput (2008) showed that those who engaged in cyber-bullying, perceived their victims as being of greater, lesser, or equal strength, and many of the victims knew their bullies in the real world (although the cyber-bullies hid their identities). It is of particular note that the students in said study indicated that the weaker victims were often also the victims of face-to-face bullying, whereas those viewed as stronger were bullied due to the anonymity that information and communication technology (ICT) affords. Along these lines, Fauman (2008) suggested that anonymity may minimize the need for cyber-bullies to be more powerful than their victims. Anonymity appears to be a key feature of cyber-bullying for those who report that they would not bully anyone offline (Vandebosch and van Cleemput 2008). This finding highlights the potential for increased cyber-bullying – given that many more people can engage in it than would normally bully anyone face-to-face.

Military services emphasize personal accountability and integrity. Thus, to the extent that social media sites provide a potential mechanism for anti-ethical behavior, such platforms can become problematic in a service academy setting.

1.5 Research Questions

In order to address the questions raised above, we examined whether military cadets and faculty members at an academy in the United States perceive social media use as affecting academic performance and behavioral conduct. Specifically, we fielded online surveys of cadets and faculty at the academy in question. In both surveys, we asked respondents to evaluate the amount of time cadets spend using social media as well as any perceived impacts on academic performance. In the cadets' survey, we also asked about the motives for using social media. Finally, we surveyed cadets on the social media sites that are most popular among them and analyzed which, if any, potential implications the use of these sites might have for behavioral conduct or the social environment.

Given the lack of prior research on social media use at military service academies, we took an exploratory approach to surveying cadets and faculty – without hypothesizing in advance whether the patterns that have emerged at civilian universities would replicate at a service academy or not.

2 Method

2.1 Participants

In the fall of 2018, academy staff used a computer-generated process to randomly select 2158 cadets for the survey. We are assuming that the list of email addresses used is accurate and current. As such, we can assume that all those who did not respond are eligible sample persons whom we are therefore considering as non-respondents. Cadets received a link to the survey via email. They had to be at least 18 years old to participate and did so on a voluntary basis. Cadets were informed that the survey is designed to examine their experience at the academy as well as the role that social media usage plays in that experience. They were told that their answers will play a significant role in helping the academy understand the relationship between social media and cadets' experiences. Finally, cadets were informed that with this understanding, the academy can take steps to enhance the cadet experience and better enable cadets to achieve their goals. Information regarding class year, gender, race, and other demographics was not collected to prevent the possibility that responses could be attributed to a specific cadet.

In addition to surveying cadets, we fielded a separate survey of the faculty, including faculty members, officers, and noncommissioned officers who have a role in training cadets. Faculty members in the Mathematics Department (47) and the Aeronautics Department (38), as well as 43 officers and 48 noncommissioned officers were contacted via email, making it a total of 176 contacts. Once again, we are assuming that the list of email addresses used is accurate and current. As such, we can assume that all those who did not respond are eligible sample persons whom we are therefore considering as non-respondents. The email sent to faculty and staff contained a link to the survey. It explained that social media use at the academy has become a topic of increasing interest to academy leadership and that, with the support of academy leadership and the Human Research Protection Committee, MITRE Corporation researchers are fielding a survey of faculty and staff on their perceptions of social media use among cadets. Within the survey, respondents were told that the survey explores faculty and staff members' perceptions of the role that social media use plays in cadets' experience at the academy. As was the case for the cadet survey, respondents were told their answers will help the academy understand the relationship between social media and cadets' experiences so that it might implement policies that enhance the cadet experience and better enable cadets to achieve their goals.

2.2 Procedure and Survey Questions

The survey resided on a server housed at MITRE, and potential respondents connected to the survey via a link, using encrypted (HTTPS) connections. The cadet survey included a variety of questions on social media use and possible impacts on performance. Given the potential for such questions to invoke socially desirable responding (Maccoby and Maccoby 1954), in which respondents give answers biased toward their perceptions of what is 'correct' or socially acceptable, we examined techniques for reducing such responding. Previous researchers have found that phrasing questions 'indirectly' (e.g., from the perspective of another person or group) can reduce socially desirable responding (Fisher 1993; Pacolet et al. 2012; and Simon and Simon 1975). As such, we phrased the questions so that cadets could answer regarding a 'typical cadet in your class' as opposed to themselves. Phrasing the questions this way was designed to take the focus off individual respondents and to elicit subjective perceptions of social media use by others.

The cadet and faculty surveys included a variety of closed and open-ended questions pertaining to social media use. Respondents were asked to provide their evaluations of the amount of time cadets spend using social media both inside and outside of class. Faculty were asked whether they had ever observed a cadet (or cadets) using email, social media, and/or the internet during a lecture. Both cadets and faculty were asked whether they believe social media usage has enhanced/had a positive impact on or hindered/had a negative impact on academic performance. In addition, cadets were asked about key motives for using social media. Finally, cadets provided their opinions as to which social media services a typical cadet in their class uses most frequently. Response options for closed survey items were presented as Likert scales to provide a standardized way of assessing behavioral and attitudinal variation on the variables of interest. It should be noted, however, that the cadet and faculty surveys were planned as part of independent studies and were not originally intended to have comparable phrasing. Although comparisons between cadet and faculty responses are provided, the slight differences in question wording limit a direct comparison between these responses.

3 Results

3.1 Response Rates

Of the 2158 cadets contacted, a total of 894 responded to the survey. Two cadets logged into the survey twice but completed it only the second time; thus, the first entries were deleted. Another two logged into the survey twice and filled it out with different responses each time. Both entries were deleted for each of these cadets, as it was impossible to determine which entry represented their 'accurate' answers. In total, we deleted six survey entries on the grounds that they were duplicates. Details on the methodology and calculation of the response rate can be found below.[1] Using the methodology described below, we determined the response rate to be 37.8%.

[1] Among the returned cadet survey questionnaires, we defined 'complete' questionnaires as those with answers to 80.1–100% of the applicable questions, 'partially-complete' questionnaires as those with answers to 50.1–80% of the applicable questions, and 'break-offs' with answers to up to 50% of the applicable questions. Using this standard, a total of 816 cadets completed the questionnaire (91.7%), a total of 26 cadets completed the questionnaire partially (2.9%), and a total of 48 cadets broke off participation (5.4%). With this information, we calculated the Response Rate 1 (from the American Association for Public Opinion Research 2016 Standard Definitions) as follows: the number of complete interviews divided by the number of interviews (complete plus partial) plus the number of non-interviews (refusal and break-off plus non-contacts plus others) plus all cases of unknown eligibility (unknown if housing unit, plus unknown, other). Specifically, the formula is: $I/[(I + P) + (R + NC + O) + (UH + UO)]$. Plugging in our numbers, the Response Rate 1 is $816/[(816 + 26) + (1264 + 48 + 4)] = 37.8\%$.

Regarding the faculty and staff members contacted, a total of 48 individuals responded to the survey. Following the methodology described below,[2] we determined the response rate to be 17.6%.

In order to put these response rates in perspective, in their respective surveys Kelty and Bierman (2013) obtained a response rate of 44%, while Elliott et al. (2011) obtained a 45% response rate. In a study by Diramio et al. (2015)…

> … a sample of 167 students was obtained via online survey from an estimated population of 1,800 veterans at seven public institutions (four research universities and three community colleges), all within a single state located in the Southeastern United States [...] This yielded a response rate of approximately 11%, which is low yet within a range typical for online surveys.

3.2 *Time Spent Using Social Media*

Paralleling previous findings among civilians (e.g., Anderson and Jiang 2018) most cadets and faculty noted that cadets spend 'more time than is ideal' or 'too much time' on social media (Figs. 1 and 2). Cadets' open-ended survey comments suggested that they are spending more time on social media than would be beneficial for their performance. Specifically, of the cadets who offered a comment one way or the other in response to the question 'How much time would you guess that a typical cadet in your class spends actively using social media?', 39 expressed an opinion suggesting that social media use is excessive, while 18 offered an opinion that it is not.

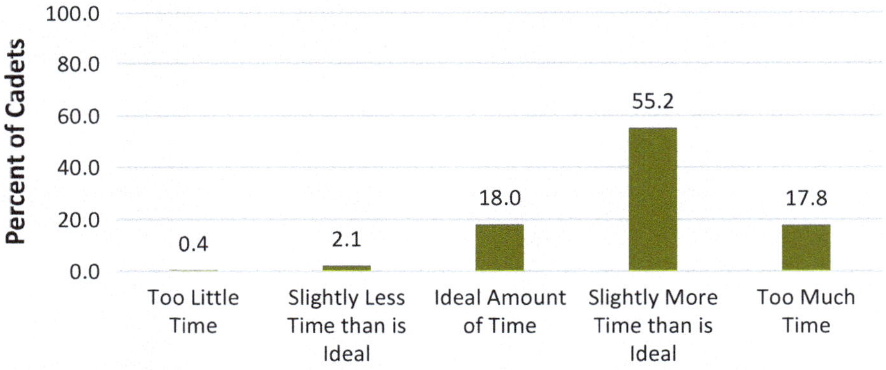

Fig. 1 Cadets' Responses to "How much time would you guess that a typical cadet in your class spends actively using social media?" (n = 890)

[2]Within the faculty/staff survey, 31 respondents completed the questionnaire (63.3%), eight respondents completed the questionnaire partially (16.3%), and nine respondents broke off participation (18.4%). With this information, we calculate the Response Rate 1 (from the American Association for Public Opinion Research 2016 Standard Definitions) as follows: 31/[(31 + 8) + (128 + 9)] = 17.6%.

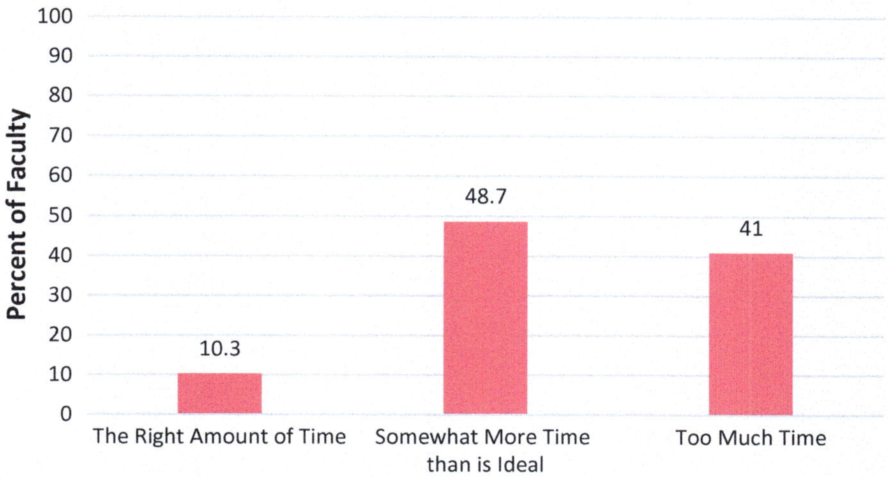

Fig. 2 Faculty Responses to "As far as you can tell, how much time do cadets spend using social media?" (n = 39)

3.3 Use of Media During Lectures

In addition to noting that most cadets spend more time than is ideal or too much time on social media, most faculty respondents had observed cadets using media during lectures (Fig. 3).

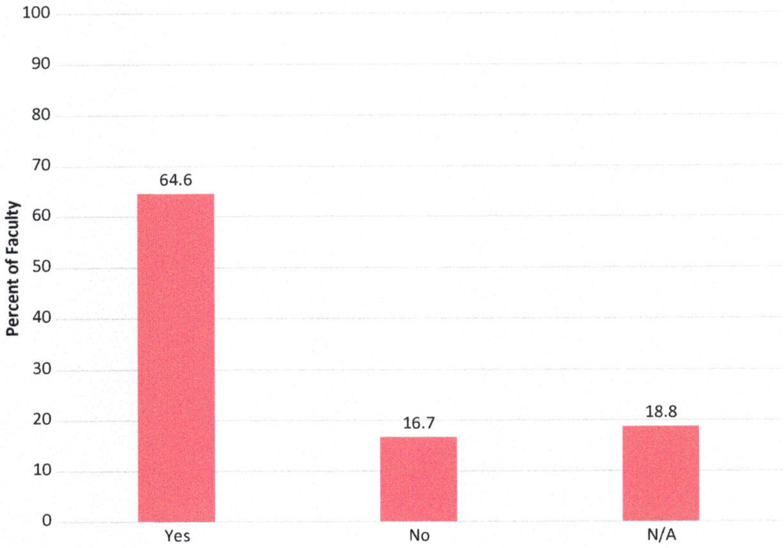

Fig. 3 Faculty Responses to "Have you ever observed a cadet (or cadets) using email, social media, and/or the internet during a lecture?" (n = 48)

3.4 Negative Reported Effects of Social Media Use on Academic Performance

Both cadets and faculty agreed that social media use affects academic performance negatively (Figs. 4 and 5). In response to the question 'Regarding a typical cadet in your class, would you guess that social media usage has enhanced or hindered that cadets' academic performance? (Other, please explain.)', 120 cadets offered open-ended comments along the theme that social media use hinders academic performance. Thirty cadets offered comments suggesting that social media use does not hinder academic performance, while 57 cadets expressed neutral opinions.

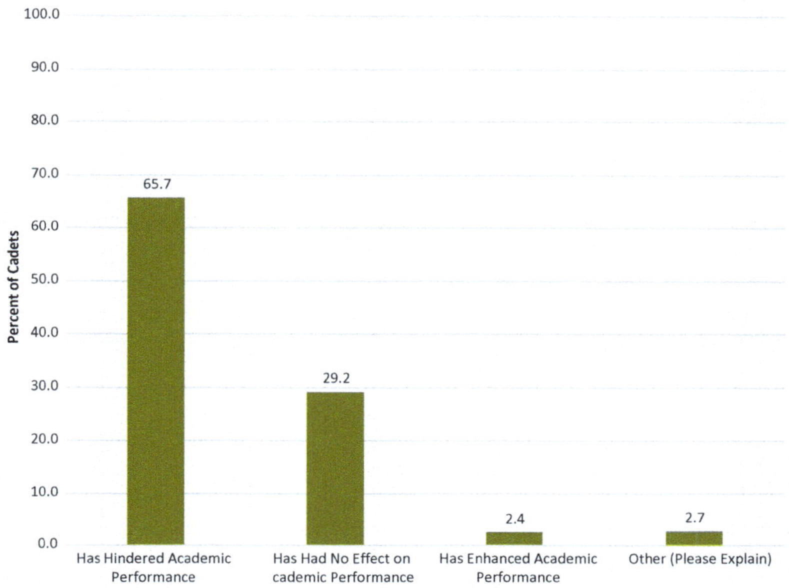

Fig. 4 Cadets' Responses to "Regarding a typical cadet in your class, would you guess that social media usage has enhanced or hindered that cadet's academic performance?" (n = 840)

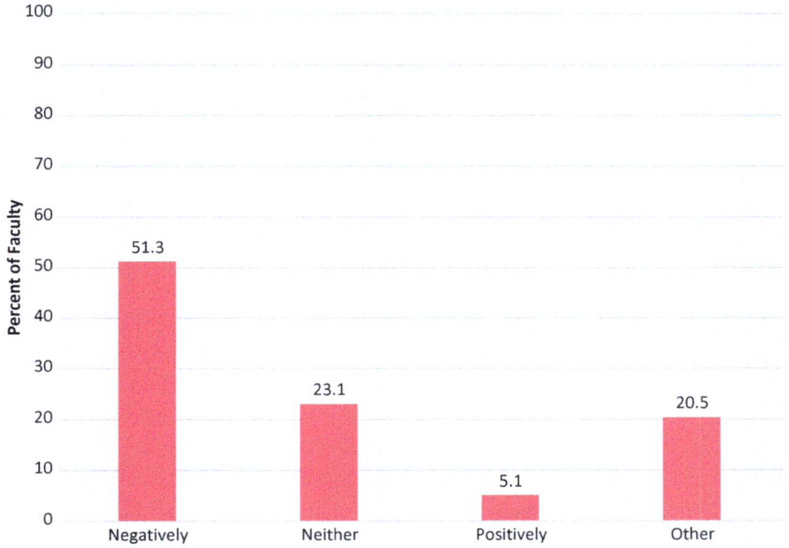

Fig. 5 Faculty Responses to "In your opinion, does social media use positively or negatively impact cadets' academic performance?" (n = 39)

3.5 Key Motive for Using Social Media: Connection

Given that both cadets and faculty agreed that social media use affects academic performance negatively, we sought to understand cadets' motives for using social media. We therefore conducted a summative content analysis (Hsieh and Shannon 2005) on the open-ended question 'If you use social media, please tell us anything you wish to tell us about what motivates you to use it.' The most common motives cadets listed included connecting with friends, family, people back home, the outside world; communication; entertainment; distraction; and boredom. In separate comments, cadets cited the isolation of academy life and the need to maintain connections outside the academy as being essential to navigating that life.

3.6 Use of Jodel

Although cadets reported maintaining connections with others as a positive use of social media, the overall behavioral picture regarding social media use was mixed. Within a different context, a study by the Pew Research Center (Anderson and Jiang 2018) found that YouTube, Instagram, Snapchat, Facebook, and Twitter (in that order) are the most popular online platforms among teens in the broader population. By contrast, cadets at the military academy we surveyed ranked the social media app "Jodel" within the top five most popular platforms, following Instagram, Snapchat, Facebook, and Twitter (Fig. 6).

These findings beg the question as to why YouTube (which 85% of American teenagers claim to use, according to the Pew study) has not attracted more of a following at the academy. A likely reason could be that cadets do not represent the general teenage population, as they are more engaged in academics and military duties. As such, the difference may be an artifact of the demographics from which the studies draw their samples.

Equally intriguing is the popularity of Jodel at the academy, seeing as Jodel did not make the top eight among American teenagers, as reported in the Pew study. Jodel allows users to maintain their anonymity and contains chat channels within a limited geolocation. Given these parameters, it is possible that Jodel appeals to those who wish to post messages on a non-attribution basis. For example, in response to the question 'Please tell us anything you wish to tell us regarding the role that social media usage plays in cadets' experience at the academy', seven faculty members cited cyber-bullying via Jodel as a concern. Of these seven, two noted that superiors – including faculty themselves – are cyber-bullied via Jodel. In response to the question 'What would you guess are the specific effects that social media usage has on the life of a typical cadet in your class?', three cadets noted the use of Jodel for spreading cynicism.

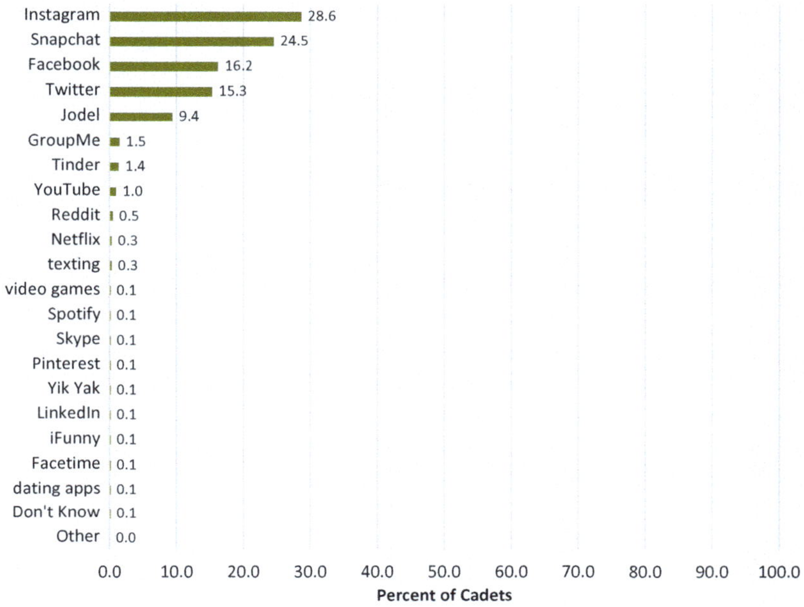

Fig. 6 Cadets' Responses to "Which social media services does a typical cadet in your class use most frequently?" (Total number of services listed was 2693. Cadets could each list up to four services.)

4 Discussion

The results of our surveys present a mixed picture of social media use at the academy we surveyed. They also suggest that, in some ways, cadets at this academy resemble their civilian counterparts in some ways, but differ from them in other ways. Similar to their civilian counterparts (whose social media use has increased dramatically over the past few years), according to most cadet and faculty respondents, cadets are perceived as spending more time on social media than is ideal. The time spent on social media has often extended into class time, as most faculty respondents had observed cadets using media during lectures. Given previous research showing that non-academic internet use during class predicts poorer academic performance (Ravizza et al. 2017), our findings regarding cadets' use of media during class hold possible implications for professional readiness. Underscoring this point, both cadets and faculty agreed that social media use affects academic performance negatively. This finding aligns with the results of numerous studies conducted at civilian universities, which (with some exceptions) show that non-academic social media use predicts poorer academic performance (e.g., Liu et al. 2017).

The influence of social media use is not entirely negative, however. According to our qualitative data analysis of comments on cadets' motives for using social media, they listed connections with friends, family, people back home, and the outside world as well as communication most frequently. Cadets also cited the isolation of academy life and the need to maintain connections outside the academy as being essential to navigating their academy experience. These findings are consistent with those of Skopp et al. (2016), who found that as long as social media is used in non-problematic ways, such media can bolster perceived social support and resilience among service members. Our findings are also consistent with those of research with civilians, showing that social media use can provide essential social capital for users (Liu et al. 2016; Liu and Brown 2014; Ellison et al. 2007). In that sense, if used in moderation, social media may have significant positive implications for cadets' psychological well-being, which can help them meet the demands of the academy while coping with the reported isolation of that life.

When social media is used in problematic ways, however, such behavior has implications that appear to be unique to military settings. Our findings showed that cadets ranked the social media platform Jodel among the top five most popular platforms at the academy, whereas civilian respondents did not rank Jodel in their top five. With Jodel's provision of anonymity, the platform undoubtedly serves as an attractive outlet for cadets wishing to express opinions that are not acceptable. Such conduct has potential implications that go beyond the effects of social media use on academic performance. In particular, conduct that entails bullying of other cadets has the unique potential to diminish both social and task cohesion among this relatively small and isolated student body, and thereby also to impede military training tasks that cadets must accomplish collaboratively. It should be remembered that all medically fit cadets who graduate from U.S. military academies are guaranteed an officer's commission in their respective service – whereas cadets in Reserve Officers' Training Corps (ROTC) programs and in officer candidate schools are competing for commissions that are not guaranteed. Thus, if there is a systemic negative effect on cohesion-related behavior at a service academy, it could potentially have broader effects on the military via this key commissioning source.

Equally troubling is the finding of cadets' cyber-bullying of faculty members, which threatens the military chain of command at the academy in question in addition to the potential for negative social-psychological effects on faculty members (Harrison 2018). The current findings parallel those of Vandebosch and van Cleemput (2008), who found that those viewed as stronger were bullied due to the anonymity provided by information and communication technology (ICT). The military emphasizes personal accountability and integrity. Thus, to the extent that anonymous platforms promote anti-ethical behavior, such platforms become problematic because they erode the respect that supports military chain of command structures. Disrespectful conduct also holds negative implications for the public image of the academy and of the cadets themselves. Finally, such conduct online could have negative implications for operational security if future service members use Jodel in an inappropriate manner off-base – given that Jodel operates using a geo-fence.

In conclusion, this work serves as an initial step toward understanding the multifaceted implications of social media use at a U.S. military academy and, as such, raises some urgent questions. One of these questions is concerned with how cadets can be enabled to reap the benefits of social media for connecting with friends, family, and others at their academy while balancing that social media use with accomplishing professional tasks. In many ways, this question can be applied to the civilian population in general as well. However, it is particularly significant in the case of service academy cadets and other service members because of the urgent nature of the tasks that military personnel must accomplish and the unique role of the military in upholding national security. Furthermore, the results of the current study raise the question whether similar findings regarding social media use apply to other service academies and to military bases – both domestic and overseas.

An additional consideration raised by this descriptive study relates to broader security issues. While local force security and larger national security issues related to social media use have been well-documented, the increased popularity of geo-local sites (e.g., Jodel) that provide anonymity for posters and the type of information that appears to be produced by these sites require careful study moving forward. As social media use occurs all over the world, there is also the question whether such findings apply to the armed forces of other countries.

Finally, the results of our work raise questions regarding what more can be done to prevent cadets and – potentially – military service members in general from engaging in cyber-bullying their peers as well as superiors and faculty members. Although policies and training curricula do exist, an examination of the extent to which cadets and faculty at military academies and service members in general are aware of extant policies related to social media use and cyber-bullying appears warranted. In addition, the current findings call for investigation into whether policies and curricula can be augmented or updated to account for the anonymous nature of some social media platforms. Any proposed changes to policies or curricula can be tested experimentally or quasi-experimentally to determine their effectiveness.

When it comes to the civilian population and especially college students, over a decade of research now exists on the implications of social media use for well-being, performance, and behavior. We hope that this contribution as well as the other contributions in this anthology serve as a call to arms for military planners to initiate further inquiry into the unique implications of social media use for the armed forces.

References

Alloway TP, Horton J, Alloway RG, Dawson C (2013) Social networking sites and cognitive abilities: do they make you smarter? Comput Educ 63:10–16

Anderson M, Jiang J (2018) Teens, social media & technology 2018. Pew Research Center, Washington, DC, p 31

Bjornsen CA, Archer KJ (2015) Relations between college students' cell phone use during class and grades. Scholarsh Teach Learn Psychol 1(4):326

Diramio D, Jarvis K, Iverson S, Seher C, Anderson R (2015) Out from the shadows: female student veterans and help-seeking. Coll Stud J 49(1):49–68. https://search.ebscohost.com/login.aspx?direct=true&db=a9h&AN=101598372&site=ehost-live

Doleck T, Lajoie S (2018) Social networking and academic performance: a review. Educ Inf Technol 23(1):435–465

Dooley JJ, Pyżalski J, Cross D (2009) Cyberbullying versus face-to-face bullying: a theoretical and conceptual review. Zeitschrift für Psychologie/J Psychol 217(4):182–188
Eagan K, Stolzenberg EB, Zimmerman HB, Aragon MC, Sayson HW, Rios-Aguilar C (2016) The American freshman: national norms fall 2016. Higher Education Research Institute, UCLA, Los Angeles
Elliott M, Gonzalez C, Larsen B (2011) U.S. military veterans transition to college: combat, PTSD, and alienation on campus. J Stud Affa Res Pract 48(3):279–296
Ellis Y, Daniels B, Jauregui A (2010) The effect of multitasking on the grade performance of business students. Res High Educ J 8(1):1–10
Ellison NB, Steinfield C, Lampe C (2007) The benefits of Facebook "friends: " social capital and college students' use of online social network sites. J Comput-Mediat Commun 12(4):1143–1168
Fauman MA (2008) Cyber-bullying: bullying in the digital age (book review). Am J Psychiatry 165:780–781
Fisher RJ (1993) Social desirability bias and the validity of indirect questioning. J Consum Res 20(2):303–315
Golub TL, Miloloža M (2010) Facebook, academic performance, multitasking and self-esteem. In: 10th special focus symposium on ICESKS: information, communication and economic sciences in the knowledge society
Harrison AM (2018) An investigation into bullying and cyberbullying: the effects of anonymity and form of bullying on severity of victim impact (Master thesis). https://etd.ohiolink.edu/pg_10?0::NO:10:P10_ACCESSION_NUM:dayton1525184128335822
Hsieh HF, Shannon SE (2005) Three approaches to qualitative content analysis. Qual Health Res 15(9):1277–1288
Judd T (2014) Making sense of multitasking: the role of Facebook. Comput Educ 70:194–202
Junco R (2012) Too much face and not enough books: the relationship between multiple indices of Facebook use and academic performance. Comput Hum Behav 28(1):187–198
Junco R, Cotten SR (2012) No A 4 U: the relationship between multitasking and academic performance. Comput Educ 59(2):505–514
Junco R (2015) Student class standing, Facebook use, and academic performance. J Appl Dev Psychol 36:18–29
Karpinski AC, Kirschner PA, Ozer I, Mellott JA, Ochwo P (2013) An exploration of social networking site use, multitasking, and academic performance among United States and European university students. Comput Hum Behav 29(3):1182–1192
Kelty R, Bierman A (2013) Ambivalence on the front lines: perceptions of contractors in Iraq and Afghanistan. Armed Forces Soc 39(1):5–27
Kirschner PA, Karpinski AC (2010) Facebook® and academic performance. Comput Hum Behav 26(6):1237–1245
Kohut A, Parker K, Keeter S, Doherty C, Dimock M (2007) How young people view their lives, futures and politics: a portrait of "Generation Next". Pew Research Center, Washington, DC
Konheim-Kalkstein YL, Stellmack MA, Shilkey ML (2008) Comparison of honor code and non-honor code classrooms at a non-honor code University. J Coll Character 9(3). https://doi.org/10.2202/1940-1639.1115
Liu D, Brown BB (2014) Self-disclosure on social networking sites, positive feedback, and social capital among Chinese college students. Comput Hum Behav 38:213–219
Liu D, Ainsworth SE, Baumeister RF (2016) A meta-analysis of social networking online and social capital. Rev Gen Psychol 20(4):369–391
Liu D, Kirschner PA, Karpinski AC (2017) A meta-analysis of the relationship of academic performance and social network site use among adolescents and young adults. Comput Hum Behav 77:148–157
Maccoby EE, Maccoby N (1954) The interview: a tool of social science. In: Handbook of social psychology, vol 1. AddisonWesley, Cambridge, MA, pp 449–487

Madden M, Jones S (2002) The internet goes to college. Pew internet & American life project, vol 15. Pew Research Center, Washington, DC
Orzech KM, Grandner MA, Roane BM, Carskadon MA (2016) Digital media use in the 2 h before bedtime is associated with sleep variables in university students. Comput Hum Behav 55:43–50
Pacolet J, Perelman S, De Wispelaere F, Schoenmaeckers J, Nisen L, Fegatilli E, Krzeslo E, De Troyer M, Merckx S (2012) Social and fiscal fraud in Belgium. A pilot study on declared and undeclared income and work. Acco, Leuven
Ravizza SM, Uitvlugt MG, Fenn KM (2017) Logged in and zoned out: how laptop internet use relates to classroom learning. Psychol Sci 28(2):171–180
Salaway G, Caruso JB, Nelson MR (2008) The ECAR study of undergraduate students and information technology, 2008 (Research Study), vol 8. EDUCAUSE. Center for Applied Research, Boulder. Retrieved 29 Dec 2010
Simon J, Simon R (1975) The effect of money incentives on family size: a hypothetical-question study. Public Opin Q 38(Winter):585–595
Skopp NA, Alexander CL, Durham T, Scott V (2016) Positive and negative aspects of Facebook use by service members during deployment to Afghanistan: associations with perceived social support. Psychol Pop Media Cult 7(4):498
The American Association for Public Opinion Research (2016) Standard definitions: final dispositions of case codes and outcome rates for surveys, 9th edn. AAPOR, Lenexa
Vandebosch H, van Cleemput K (2008) Defining cyberbullying: a qualitative research into the perceptions of youngsters. Cyberpsychol Behav 11:499–503
Xanidis N, Brignell CM (2016) The association between the use of social network sites, sleep quality and cognitive function during the day. Comput Hum Behav 55:121–126

The Need for Visibility: The Influence of Social Media Communication on Swiss Armed Forces Officers

Eva Moehlecke de Baseggio

Abstract Social media is becoming increasingly central within armed forces' communication strategies. While there is research on the external impacts of social media communication, less is known about the effects on actual employees. Thus, the focus of this contribution is to collect and analyse the wishes, needs, and attitudes of officers of the Swiss Armed Forces that relate to social media communication in particular and communication in general. Designed as an explorative study, 34 semi-structured interviews were conducted and a qualitative content analysis was carried out. Besides some rather instrumental issues related to social media communication, three soft factors related to and affected by (social media) communication emerged. These are visibility, identification, and commitment. The three concepts are interrelated and at least partly dependent on each other. The analysis of the interviews and the theoretical embedding show the high sensitivity of cadre members towards organisational communication as well as the many ways in which communication affects their work, motivation, and well-being.

Keywords Social media · Semi-structured interviews · Content analysis · Visibility · Organisational identity · Affective commitment

1 Introduction

The Swiss Armed Forces, hereinafter referred to as SAF, regard social media mainly as a tool for reaching out to society and especially to young people, as they are the ones forming the recruitment pool for the armed forces. However, in 2016, SAF leadership also identified the cadre personnel of the SAF as an important stakeholder group for the SAF's official social media communication. The starting point

---

E. Moehlecke de Baseggio (✉)
Military Academy (MILAC) at ETH Zurich, Birmensdorf, Switzerland
e-mail: eva.moehlecke@moehlecke.com

E. Moehlecke de Baseggio et al. (eds.), *Social Media and the Armed Forces*, Advanced Sciences and Technologies for Security Applications,
https://doi.org/10.1007/978-3-030-47511-6_4

for this was the need to improve cadre recruitment in order to satisfy SAF staffing requirements, as sufficient and stable cadre personnel numbers are decisive for the SAF to successfully fulfil their mandates. A catalogue of measures was defined to master the staffing challenge. Although the information and communication needs of cadre personnel were part of this package, there was a lack of a deeper understanding and empirical evidence. Therefore, research concerning communication needs and desires of cadre employees was necessary.

At the same time, the SAF command and its communication department chose to pay special attention to the contemporary social media platforms related both to reaching out to youth and the recruitment[1] of cadre personnel. Social media was new to the SAF, and again, scientific data and knowledge about the effects of social media communication were welcome to gain insights into the influence these platforms have on the SAF and on Swiss society. Thus, in order to combine both requirements, an extensive research project was launched, which also included the analysis of the specific needs and desires of cadre personnel related to social media. However, as it is difficult to collect data on the needs and desires connected to social media alone while leaving general communication needs aside, the cadre subproject comprises both communication needs and attitudes in general as well as beliefs and desires regarding social media communication in particular. The nature of the project is explorative, which gave leeway for qualitative research in the form of semi-structured interviews.[2]

The project being explorative, the state of research and most of all the epistemic interest defined the research topic, which in turn defined the interview questionnaire. Thus, the state of research and the research question pursued are described in the following, before the method applied as well as the results including the theoretical formulation of the findings are explained and discussed.

2 State of Research and Research Question

Social media communication has become a firm component in the external communication of organisations. In a survey among 2710 communications experts of joint-stock or private enterprises, public sector agencies, and non-profit-organisations across Europe in 2016, the respondents rated social media as the third most important channel and predicted it to be the second most important channel by 2019 (Zerfass et al. 2016). Some academic research is being conducted on Public Relations, focusing mainly on the interrelations between organisations and their stakeholders and reputation management (Pleil and Matthias 2017). Three out of four organisational communication experts believe that social media impacts and

[1] The SAF are based on a militia system and on conscription and thus do not need to recruit in the original sense.

[2] The author wishes to thank Jennifer Victoria Scurrell, who in large parts conducted and analysed the interviews on which this article is based.

changes stakeholders' perception of an organisation (ibid.). Among the stakeholders are the employees of an organisation. Under the term of employer branding, organisations aim to strengthen staff retention or to recruit new employees (Karnica and Kumar 2019; Pleil and Matthias 2017). Social media has become a relevant element regarding recruitment. In their study among MBA students, Karnica and Kumar (2019) were able to confirm a moderating effect of social media on the person-organisation fit – which stands for the compatibility of a person's values with those of the organisation – and the evaluation of an organisation as an attractive employer.

Social media thus enables organisations to intensify the public awareness of their brand, including their employer brand. This effect goes back to the freely available information about an organisation on social media, which is in line with the findings that visibility is one of the four central affordances of social media (Treem and Leonardi 2013). The effects of social media on organisations are also the focus of corporate identity scholars (Devereux et al. 2017). The social media presence of organisations contributes to heightened visibility and intensified relationships with stakeholders (ibid.). Due to the fact that this type of media is designed for interaction, it inherently bears the technological prerequisites to start a dialogue or interact with the conversation partners. Thereby, social media potentially establishes relationships between the communicating parties (Kissel and Büttgen 2015). However, without a clear communication concept, including an organisation's communication targets and the stakeholders' social media behaviour, needs and expectations, social media communication will not succeed (Pleil and Matthias 2017). Furthermore, the interrelations between communication and employee engagement have been researched for communication in general, but not specifically for social media communication. A beneficial internal communication contributes to increased employee engagement by providing employees with the resources necessary to fulfil their work, and beyond that also deepens workplace relationships (Karanges et al. 2015). The external communication of an organisation, in turn, is found to be linked with employee productivity, job satisfaction, and trust (Kandlousi et al. 2010).

Thus, the goal of the interviews and the subsequent analyses is to answer the following research questions: how does the social media communication of the SAF affect their cadre personnel? Can the SAF's social media communication contribute to the recruitment and retention of cadre members, and if yes, by which mechanisms?

3 Method

Data collection took place by conducting a total of 34 interviews, 31 of which were semi-structured. Furthermore, one focus group interview with four Press and Information Officers (PIOs) was conducted, as well as two expert interviews with communications experts, all of whom are cadre members of the SAF themselves. Semi-structured, guided interviews are an ideal method for the present research questions. On the one hand, they offer the chance to touch upon all relevant issues, while, on the other hand, leaving enough room to capture the wider context of meaning of the interviewees' statements (Rager et al. 1999). Additionally, semi-structured interviews allow researchers to capture the ideas of a small population group, such as the SAF's cadre personnel, and to perform explorative analyses (ibid.). They further not only ensure a retrospective assessment of the phenomena associated with the research question pursued, but also include a real-time evaluation (Gioia et al. 2012).

The interviews were conducted between February and June 2018. The sample for the 37 interviewees was set up according to theoretical considerations and involved both professional and militia cadre personnel, whereby 21 of the participants are professional officers (or aspirants for professional officers), and 16 are conscript officers. The SAF are based on conscription and a militia system. Conscription applies to young men only, whereas women can participate on a voluntary basis. The constitution explicitly does not allow a regular, i.e., professional force (Federal Constitution of the Swiss Confederation 1999). Hence, the SAF consist mainly of militia troops with only a low share of professional officers.

The lowest ranked person interviewed is a Chief Sergeant Major (CSM), while the highest ranked person is the Chief of the Armed Forces, who holds the rank of Lieutenant General. In addition to the Chief of the Armed Forces, interviews with three other members of the SAF command were conducted. Of the 37 participants, six are senior staff officers. Full anonymity was granted to all participants – with the exception of the members of the SAF command and the communication experts, whom it is rather easy to single out. Therefore, anonymisation would have been of little use in those cases.

The shortest interview lasted 26:13 min, while the longest one took 1 h 17:04 min, with the average length being 46:48 min per interview. The interviews were conducted using a thematically oriented questionnaire containing questions on the general needs and wishes of cadre personnel related to communication, and on the communication channels they assess as adequate. The questionnaire also focusses on the advantages and disadvantages of social media, for which purposes cadre personnel think it is appropriate, and whether social media and the SAF match. Finally, it asks about the effects of social media on the motivation of cadre personnel, why this type of media can be motivating or, if the contrary applies, why not. Personal attitude regarding the use of social media is the subject of the next part of the interviews, which includes questions about what cadre personnel would like to be integrated or changed in the SAF's social media communication, as well as questions on their private and military social media use. Apart from the questions on communication and social media, the interviewees were asked about their military ambitions as well as given a few demographic questions on, for instance, education, age or military rank. Except for the demographic ones, all questions are open and non-leading, which corresponds to methodological best practice (Gioia et al. 2012). In order for the interviewees to describe their views as freely as possible, the questionnaire served as an orientation guide rather than dictating a rigid order. Moreover, new questions, which were not included in the questionnaire, came up in the course of the interviews (Flick 2004).

After being fully transcribed, all interviews were first subjected to an analysis based on the method of grounded theory by Strauss (2004). Performing a text-based analysis allows to go beyond the manifest text, and to include a further level of meaning which may not be apparent at first glance (Mayring 1994). Next, a qualitative content analysis was conducted, which in a first step consisted of open coding using inductive in vivo codes (Böhm 2007). Three randomly chosen interviews were independently in vivo coded by the author and another researcher. The results were compared and grouped into categories – the axial codes (ibid.) – which were formulated based both on empirical evidence and theory. For the next two interviews, which were also randomly chosen, the content analysis was again carried out doubly by both researchers. One person did the analysis using the draft category system consisting of the axial codes, while the other used open coding. Finally, the results were compared, and the draft category system was refined. The remaining interviews, except for those with the members of the SAF command, were subjected to a content analysis using the refined category system, which was constantly vetted. In a third and final step, the selective coding, the central and integrating phenomenon of 'communication' emerged. This motive seems obvious and was already present in the research question, which, according to Böhm (2007), is a common thing to happen. The goal of this type of analysis is to condense the extensive raw material of the interview transcripts into common underlying motives shared by the participants. Thus, the emerging category system can be understood as a translation of the interview texts into generalised statements, whereby the term generalisation accounts for the analysed group only.

The four interviews with the SAF command were also analysed separately by both researchers by means of open coding. For each of these interviews, a separate category system was set up, and the eight resulting category systems were compared and analysed.

As a generalised statement, the needs, desires, opinions, and attitudes of the SAF's cadre personnel can be summed up into the main axial categories, *visibility*, *identification*, and *commitment*. The categories do not stand isolated from each other, but are interrelated and show high degrees of interdependencies. Table 1 shows the categories as well as some corresponding in vivo codes.

The results related to the main axial categories, which in the process of forming them were grounded on inductively collected in vivo codes and theoretical, hence deductive formulation, will be presented and theoretically explained in the following. The methodological procedure was based on a mixed inductive and deductive approach (Gioia et al. 2012; Mayring 2000). Firstly, however, some specific statements of cadre members regarding social media as communication channels of the SAF will be presented and complemented with the views of the communications experts.

Table 1 Overview of content analysis main categories and extract of in vivo codes

Selective category	Axial category	In vivo code (extracts only)
Communication	Visibility	Do good and talk about it.
		Need for more proactive communication
		Society does not know what we do.
	Identification	Better information flow would increase identification with the SAF.
		Need for a feeling of belonging to the SAF/a subgroup in the SAF
		Unsystematic social media communication
	Commitment	The SAF is a matter close to the heart.
		Proud of what we do.
		Militia cadre is not involved enough in the information flow.

4 Social Media as Communication Channels of the SAF

The SAF's cadre personnel almost unanimously understand social media as simply another instrument in the communications tool kit. However, it is a relevant one, seeing as almost a third (32%) of the Swiss electorate knows at least one of the SAF's social media channels (Szvircsev Tresch et al. 2019). Thus, most of the participants assess social media as a necessity and an accompanying tool to the more traditional communication channels. They further recognise the potential of social media to reach youth and think of it as a means to "create a positive noise level" (Interview 16).[3]

When asked about the communication purposes for which cadre personnel think social media is adequate, they cite the possibility to tell positive stories in a modern, proactive way, to use it for Public Relations, and to motivate young people to do military service. They associate these purposes with one of the advantages inherent in social media platforms, namely the chance to communicate through pictures and videos. These visual media types are a good means to transport and evoke emotions, something that is much harder to achieve with words only (Fahmy et al. 2006). Most of the participants do not see a field of tensions or even conflicts between the open and informal social media platforms and the hierarchically organised, rather closed structure of the SAF. When asked whether the SAF and social media match, one interviewee puts it as follows:

> In my eyes, they absolutely match. Because our recruits, the present ones and our future recruits, those are the ones that are on these platforms. And if [the platforms] don't fit the SAF, then we are just outdated in that area. Then it's our problem and not the problem of the young ones, that's my opinion. (Interview 11)

[3] All quotes taken from the interviews were translated from German into English by the author.

The expert and focus group interviews with the communications experts – all of them are cadre personnel of the SAF, as well – draw a very similar picture. In the experts' views, the most central advantage of social media is its immediacy and seemingly low costs compared to print products or adverts. They endorse the opinion of the above quoted interviewee that social media is an adequate means to reach youth. On social media, it is possible to show young people that the military service can be interesting and enjoyable. In addition to these instrumental characteristics, the experts attribute another dimension of meaning to social media platforms, which is the possibility to increase the self-esteem of the SAF's personnel by giving them a stage to present what they actually do in the SAF. Consequently, the communication experts assess the SAF's personnel as well as their relatives and friends as the main target group for the SAF's Facebook presence. They also list social media as a way to show the true face of the SAF, and to inform the (civil) population about what services the SAF do for society, to endow sense, and to communicate in a transparent, open, and authentic way. Thus, to the experts, the goal of such social media communication is to generate credibility.

In the interviews, the experts touch upon soft-factor effects of social media communication on the SAF's cadre personnel, such as a means to increase self-esteem. These factors gain considerable weight in the interviews with the other cadre members, who do not work for the communication department and hence possibly have a less instrumental perspective on social media communication. The factors are reflected in the three axial categories already mentioned above: visibility, identification, and commitment. The interrelations between the SAF's communication and these aspects will be discussed in the following, with specific focus on the matters related to or reinforced by social media.

5 Visibility: The Source of Appreciation and Respect

The interviewed SAF cadre members consistently stated a desire for the SAF's official communication to be more proactive. They wish for communication to be positive, transparent, credible, and authentic, or in the words of one participant:

> On the contrary, the army should very much present itself, again the word proactive, as modern, open, you know. What cannot happen is that the population gets the impression that there are just order-receiving steel helmets here, who somehow mourn the Cold War and still prepare for it. This cannot be, rather the army should present itself as a modern, lean instrument, in which one can profit a lot for civilian life, for oneself. (Interview 15)

Transparency, credibility, and authenticity imply the notion to portray the organisation's complexity and diversity – the latter referring to units, functions, and tasks rather than demographic diversity. By communicating more proactively, the SAF cadre members hope to increase the external visibility of the SAF and thus to achieve a higher degree of information in society on what services the SAF offer and how taxpayers' money is spent. This is done – consciously or unconsciously – with the aim to legitimate the organisation and, associated therewith, to obtain public appreciation.

Conversely, information is a predecessor for the formation of external appreciation. Society needs to be informed on the missions and performance of the SAF, and cadre members are aware of that. This probably rather intuitive assessment of the impact that communication can have is supported by a study conducted by Ho and Cho (2016). By analysing the perceived performance of the police, the study found that in situations with a lack of information, people base their assessment on preconceptions and clichés, which generally leads to a lower assessment of the performance, thus resulting in lower trust invested in the police (ibid.). On the other hand, when people perceive the police as a communicating agency, they judge their performance higher and invest higher trust in them, independent of the actual performance (ibid.). In the case of the SAF, the participants feel that the population is ill-informed about the missions the SAF need to fulfil and what military service looks like. One interviewee expresses this opinion as follows:

> And I just see that there are many people, of all categories, from ten to 70, 80 years old, they just simply don't know what we do. And that's really a pity, because we do good things. We don't sell, well, not sell, but we just don't show it enough or did not show enough, that we do this. (Interview 3)

Another participant uses even more pointed words by stating that "the army has zero visibility, and thus zero support in the population. And that's the army's doom, and we cannot steer towards that. The army has to be perceived by the public" (Interview 37).

Cadre personnel feel that communicative visibility would support them in gaining the much-needed public appreciation, which in their view has faded along with the decreasing societal significance of the SAF. They explicitly state the belief that if the SAF communicate extensively, appreciation will return, both externally in society and internally in the organisation, as the SAF and their actions will be more visible again. One officer tells a story about how he organised a parade through a village with his troops.[4] The soldiers' motivation for the parade was low and at a certain point, he himself started to doubt the idea, especially as it required great efforts. Nevertheless, he decided to go ahead with the parade. He recounts:

> There were a lot of people standing on the side of the road. And they applauded when the troops came by. … Yes, there were so many people, it was unbelievable. The effect it had! The soldiers themselves and the officers too, I noticed afterwards, they almost were like 5 to 10 centimetres taller afterwards. They had never seen that before, that there are people at the side of the road who applaud. (Interview 37)

---

[4] In the Swiss context, military parades are very unusual. The last major military parade in Switzerland took place 30 years ago (see Aargauer Zeitung (2019): Why There Have Been No Military Parades in Switzerland for a Long Time [online]. [Viewed 18 February 2020]. Available from: https://www.aargauerzeitung.ch/schweiz/warum-es-in-der-schweiz-schon-lange-keine-militaerparaden-mehr-gibt-134741044)

As the troops cannot constantly parade in the streets of Switzerland, social media is, in the eyes of one participant, an ideal means to get a feeling for the appreciation the public attributes to the SAF and their personnel. The interactivity of social media platforms allows those who like to do so to comment on what they have just read, heard or seen – even if it is just a like or an emoji that they post as a comment. It lends a feeling of respect and recognition when people react to the SAF's social media posts with comments such as "Hey, great stuff!", or "We are proud of what you do, boys!" (Interview 36). Accordingly, the possibility of making one's work visible on social media, or even to generally increase the organisation's visibility, is one of the main affordances that make social media so attractive to organisations (Treem and Leonardi 2013).

However, it is not just the organisation which needs acceptance. Its employees and cadre members have a similar need to feel accepted and appreciated. Without an informed public, these needs cannot be met. On the contrary, the lack of information mentioned by cadre members leads to the converse feeling of not being appreciated by society for their work, as states one participant:

> Somehow, I have the impression that the army doesn't do enough to make the population aware of what we effectively do for them, like what would go wrong if the army didn't exist anymore. Many organisations would be lost if the army couldn't show up anymore with all the means it has. (Interview 30)

Thus, visibility and positive communication contents are of the utmost importance for the SAF and their cadre personnel, given that the interviewees link external visibility with appreciation. Being represented in the SAF's communication makes them feel seen and appreciated as individuals. This desire is not unusual – people have the desire to make their work seen by others, as it lends them a sense of empowerment by feeling visible and thus respected and appreciated (Baroncelli and Freitas 2011; Boons et al. 2015; Suchman 1995). Fittingly, the participants identify social media platforms as the adequate tools to boost the visibility of their organisation, including its different units, members, and the corresponding functions.

Inner-organisational appreciation is another important subject linked to visibility. Cadre personnel are very sensitive with regard to who is portrayed how in the SAF's communications products. They are deeply perceptive of the unequal featuring of commands, divisions, regiments, units, and military functions, particularly on social media. The conception that one's own unit is under-represented is unanimously shared. One participant stated a certain lack of understanding of the fact that the graduation ceremony of one of the training courses was portrayed, whereas the graduation ceremony of his own training course was not. Further comments in this regard were frequently made in the interviews. Similarly, many feel that they only see armoured infantry and grenadiers in the SAF's communication, especially on social media channels addressed to youth, and criticise that there is more to the SAF than just jets and tanks. Logistics, transport or other back office units are rarely portrayed, which the participants interpret as not being attractive enough. They often consider this unequal communication behaviour to be a lack of inner-organisational appreciation for the members of the remaining units.

Cadre personnel also feel that an increased visibility of the SAF via social media would make their professional lives easier. Better informed recruits, for example, have a more realistic idea of the SAF, which translates into more realistic expectations. It goes without saying that such recruits are easier to train. Thus, as was mentioned above, social media platforms are the ideal channels to address and inform future recruits in an adequate way.

6 Communication and Organisational Identification

Identification comprises the classification of oneself into a group of similar people, whereby the similarity can be based on individual preferences, common activities or similar people (He and Brown 2013; Kim et al. 2010). Organisational identification of the SAF's personnel with the organisation involves shared organisational attitudes and behaviour, thereby increasing the willingness of the individual to accept inconveniences, stress, and even harm on behalf of the greater good, which in this case is the good of the SAF (Kim et al. 2010). Moreover, it relates to organisational membership (Scott 2007), and is the psychological tie that connects the organisation and its employees (Wiesenfeld et al. 1999). The stronger the organisational identification of an employee, the less the cognitive distance between oneself and the organisation is felt (Shamir and Kark 2004).

A vast majority of the interviewees looks for connection points they can use to identify more strongly with the SAF. However, although they do feel the need to identify with the organisation and to feel loyalty for it, they do not have enough obvious contact points. A career officer who expresses his need to identify more strongly with the SAF sees potential for enhancing organisational identification in web-based communication platforms. He explains his idea of a closed internet forum for professional SAF officers, believing that "the feeling of belonging would increase, like 'we are military professionals'. That could be something new, something nice." (Interview 31).

Those interviewees who feel a strong organisational identification with the SAF present a common pattern. Their value systems are rather traditionalist and conservative. As connecting points to identify with the SAF, they mention aspects such as manliness, patriotism, homeland, and the duty to serve it. Remarkably, these persons generally perceive social media as a threat rather than an opportunity, and evaluate these platforms as something which "does not stand for the values of the SAF" (Interview 4) or, as one participant puts it, they simply "don't have such a good feeling about that stuff." (Interview 14).

The majority, however, wishes for communication to be more proactive. This desire implies that cadre members evaluated the SAF's communication – such as it was at the time the interviews were conducted – as not being active enough. It is a negative assessment of at least one aspect of the SAF's communication and thus influences the organisational identification of the cadre personnel. Further criticism of the SAF's communication comes from conscript officers, who consistently state that the information flow for conscript cadre members is poor. When asked whether this makes them feel taken less seriously, one participant responds:

> "I think so, or, you know, it would just make it easier, I think, it would increase the identification with the organisation, you see." (Interview 7)

Organisational identification is entangled with communication from and with top management, so-called vertical communication (Bartels et al. 2010). The importance of vertical communication lies in promoting clarity about the organisation and its values towards the employees, and thereby in minimising possible uncertainties. The degree of identification of employees with their organisation is connected with how they value that communication (ibid.). Thus, if cadre personnel think they receive either too little or too much information from the SAF command, both is perceived as a lack of appreciation, which affects their organisational identification. Beyond that, communication is linked with organisational identification through its power to create shared meanings and thus a common context (Wiesenfeld et al. 1999). Over time, this mechanism produces an image or an identity of the organisation to which its personnel can relate, hence deepening their identification with the organisation (ibid.).

Enhanced visibility does not only affect the public's appreciation of the SAF and their personnel, but also influences the identification of the employees with the organisation itself. Seeing and recognising oneself or one's organisation in communications products creates pride and the possibility to show what cadre members actually do. Additionally, the feeling that one is being seen and respected is a precursor to organisational identification (Boons et al. 2015). Moreover, the employees' perception of their organisation as being esteemed is also an antecedent for developing organisational identification (He and Brown 2013; Jones and Volpe 2010). Through the prestige of the organisation, the individual employee can link his or her identity to that of the organisation and, by being a part of it, profit from its high reputation (He and Brown 2013; Jones and Volpe 2010; Kim et al. 2010). The external view of an organisation is in large parts, albeit not exclusively, shaped by the organisation's communication. Due to the speed, immediacy, and low costs social media entails (excluding payroll costs, of course, which cannot be underestimated), much of the SAF's day-to-day communication takes place on their social media channels. Therefore, it is of fundamental importance that these platforms be maintained at a high-quality level. Only then can social media offer first hand impressions of the public's opinion of the SAF.

Furthermore, the interactive structure of social media deepens the process of cadre members' organisational identification by reinforcing it with each interaction (Jones and Volpe 2010). Identification thus emerges in the continuous interplay of how others perceive the organisation and incorporating this perception with that of the organisation's members (Mujib 2017). Having said this, most of the interviewees stated the feeling that the SAF's social media communication seemed to be somewhat without concept. The participants think it could be much more successful if it were conceptually embedded in the organisation's integrated communication. They thereby express criticism of the SAF's vertical communication, which is linked directly to organisational identification. This criticism is further illustrated by statements referring, for instance, to the fact that the participants miss a common terminology for certain issues, which they think should be distributed to all staff members of the same ranks. Similarly, if basically necessary information is communicated in a hesitant or even secretive way, they associate this manner of communication with low appreciation and deficient vertical communication. This, in the medium and long term, can lead to a lower level of organisational identification. One interviewee explains that instead of focusing on informing people so that everybody follows the same goals, as is the case in the private business sector, in the SAF the focus lies on protecting information. According to him, the reason for this behaviour lies in the fear of the SAF that the media could interpret something in an unfavourable light, or that they might publish something the SAF does not wish to be published (Interview 7). This approach is contrary to the communication culture on social media, which is characterised by openness and informality. What is more, it stands in contrast to research studying the link between transparency and trust in governments, which found that transparency, even when something negative is reported, does not impact trust negatively (Grimmelikhuijsen 2012).

7 Affective Commitment and Motivation

Identification stands for the sense of membership to a group, which in the specific case of organisational identification is the membership to organisation-related groups. It is a rather self-referential evaluation that serves the definition of the self as a function of the group membership based on the perceived similarity in qualities or fate with the other members (Gautam et al. 2004; Jones and Volpe 2010). Organisational commitment, on the other hand, indicates a stable and enduring attitude towards the organisation, which builds on the relationship between the organisation and the individual (Gautam et al. 2004; Mercurio 2015). Although it is a multi-dimensional concept, its core essence is found to be its most emotional element: affective commitment (Mercurio 2015). Affective commitment includes feeling proud to be a part of the organisation, having an emotional bond with it, being involved, and having a sense of belonging (Felfe et al. 2006), which results in a certain attitude towards the job (Camilleri and Van der Heijden 2007; Mercurio 2015). In this context, one participant refers to the armed forces as being “a matter close to the heart”, which suggests a strong affective commitment (Interview 8).

Although there is some overlap between organisational identification and affective commitment, both concepts need to be kept separately (Gautam et al. 2004), as organisational identification actually impacts affective commitment, rather than being the same (Jones and Volpe 2010). Organisational commitment is of the utmost importance for the organisation, as it influences the employees’ performance and motivation positively (Camilleri and Van der Heijden 2007). Employees need to know what happens inside the organisation in order to develop a sense of commitment. Thus, communication plays an important role. The link between communication and organisational commitment, or more specifically affective commitment, is produced by communication satisfaction. The latter stands for the degree of employees' satisfaction with the organisation's information and communication as a whole (Chan and Lai 2017). Communication satisfaction – also a multi-dimensional concept – includes communication climate, supervisory communication, organisational integration, media quality, and personal feedback, as well as horizontal, corporate, and subordinate communication (ibid.). Communication satisfaction is an antecedent of organisational commitment: if employees are satisfied with communication in their workplace, their organisational commitment will increase (ibid.).

In this context, especially the effects of communication satisfaction on job performance need to be highlighted (Kandlousi et al. 2010). At the time of the interviews, the cadre members who participated in this study, did not feel satisfied with the SAF's communication, nor did they feel supported enough in terms of communication. Many conscript cadre members stated the wish for communicative support with regard to being able to show their families and civilian employers the usefulness and added value of their military service. They clearly sense a certain pressure to justify their absences in their civilian lives, and feel it would make it easier for them if they were able to show how skills and abilities acquired during military service can be transferred into their civilian lives. While these participants recognise the limits of the more traditional communication channels, they assess social media as the channel ideally suited to fulfil this desire.

Hence, the SAF should focus on improving communication satisfaction and consequently the commitment of their cadre personnel. Cadre members can be considered to be the soul of the organisation, meaning they are the most valuable to an organisation. Therefore, they should ideally have a strong feeling of affective commitment. However, especially conscript cadre personnel feel they are being strongly neglected in terms of communication and information. One interviewee points out that, "as militia cadre, you really have the same level of information as the public. Now for, let's say for overall subjects, you know, or things where you're directly involved, yes, but as militia cadre you don't have any head start on information" (Interview 7). The interviewee goes on to describe the level of information of a conscript cadre member as being the same level that people have at the regular's table of the local pub. Another participant connects this feeling of neglect to the elitism within the SAF, stating that "one would have to communicate more proactively. Maybe also about conscript cadre members who are close and anchored there, and not only from the high horse, from the career military cadre point of view" (Interview 8).

These quotes illustrate the conscript cadre members' sense of being excluded and not being taken seriously, which leads to feelings of a lack of organisational support and hence frustration. These feelings are detrimental to the affective commitment cadre members feel towards the organisation, as perceived organisational support is a relevant influence on affective commitment (Mercurio 2015). Another example for this mechanism is the feeling of not being visible, or the feeling of being part of an organisation that serves the public, but about which the public knows very little at times (see chapter "Visibility"). This is reflected in the following statement, in which the participant explains how he felt when he had to leave work for a few weeks of military service:

> ... [S]omebody wishes me a nice vacation, and then I sleep four hours a night and work a lot. Sure, you have to take this up to a certain degree. And I always say, you are welcome to join me for a day, but you have to do a full day so you can see how it works. (Interview 15)

Such misconceptions by the public thus show that the desired visibility discussed in the above chapter is entangled with the level of affective commitment cadre members feel. Without visibility, the external image and prestige of the organisation will suffer. Accordingly, the individual will not feel proud to be part of the organisation to the same extent as before, hence the level of affective commitment might decrease. The above statement describes a negative direction of these interdependencies. However, if they occur in a positive form, cadre personnel's motivation and their affective commitment will be strengthened.

8 Conclusion

The aim of the present analysis was to answer the research question as to how the SAF's social media communication influences their cadre members, and whether it contributes to the recruitment as well as retention of cadre members. The content analysis of the interviews conducted reveals some major effects of communication on cadre personnel. These are: the feeling of appreciation and respect through an increased visibility of the organisation, including its members and their work; the issue of organisational identification, which stands for feeling that one is a member of the organisation; and the commitment of cadre members towards the SAF, including pride in being a part of the organisation and experiencing a sense of belonging.

Cadre personnel, just as all employees, have a need to feel appreciated by their employer. In the case of the SAF, which are part of the Swiss governmental apparatus, this extends to the general public as their ultimate funders. Without knowledge of the activities of the armed forces as part of the governmental apparatus, members of the public will not invest trust in them and thus will not show respect (Grimmelikhuijsen 2012). At the same time, by communicating and informing the public about the SAF, their members and their work, the employees feel seen and appreciated by the organisation itself. Increased visibility has been shown to be one of the main advantages social media offers (Treem and Leonardi 2013). This type of media is excellently suited for informing the public and, at the same time, lending a feeling of appreciation to the SAF's employees, as social media (potentially) reaches a broad audience on both the internal and external level of the organisation. Appreciation is not a luxury good which can be spared. On the contrary, empirical evidence points to it being the most important of the three rewarding factors, financial rewards, job security, and appreciation (Stocker et al. 2010). Feeling appreciated is linked with job satisfaction and, even more so, mitigates negative effects of job-related inconveniences (ibid.).

The second issue, the organisational identification of cadre members with the SAF, is of fundamental importance. Employees who identify strongly with the organisation invest more time and effort into the organisation, show stronger motivation, productivity, and higher job satisfaction, and there is a higher probability that they remain with the organisation (Gautam et al. 2004; Kim et al. 2010; Wiesenfeld et al. 1999). Most importantly, the likelihood of employees who identify strongly with their organisation to understand the organisation's perspective and to act in its best interests is higher than it is with employees who do not identify with the organisation (Kim et al. 2010). Social media platforms foster and intensify organisational identification by shaping – together with other communication channels – the external image of the organisation. They thus create a certain image or prestige of the organisation, which boosts the identification of its employees by being part of a prestigious organisation. Moreover, the interactional character of social media potentially contributes to deepening the organisational identification with each positive interaction. Devereux et al. (2017) established how social media contributes to intensifying the relationship with the different stakeholders involved, such as the SAF's cadre members. The allegation by cadre members that the SAF's social media communication seemed somewhat without concept is a phenomenon often found in the early stages of social media communication (Pleil and Matthias 2017). The SAF reacted to this problem and put a fair amount of effort into conceptualising their social media communication, however, the interviews took place before that.

Finally, the affective commitment of the SAF's cadre members is connected to their communication satisfaction, which of course interrelates with their wish for communication to be more proactive, hence creating an increased visibility of the SAF. As was stated above, since the interviews were conducted, the SAF's communication has been reviewed, which might have considerably affected cadre member's communication satisfaction as well. One way or another, the effects of communication on the commitment of the cadre members need to be considered, as they affect all organisation members, including cadre.

The three issues elaborated in this chapter are not solely dependent upon communication, let alone upon social media communication. Although the latter can hardly be analysed separately from communication in general, it nevertheless offers a few unique characteristics, such as the vicinity to the public and access to the immediate reactions of society members as well as the interactional design of social media platforms. In order for interaction to get off the ground, the SAF should invest in managing their community and cultivate interaction, keeping in mind that the community's reactions contribute to the SAF's cadre members' feeling appreciated. Moreover, the use of pictures and videos as well as the diversity of communication content that can be published allow the organisation to portray itself in all its complexity – including the different units, divisions, and functions – and hence to work on the one aspect that lies above all others: increasing the SAF's visibility. By doing that, the concerns, desires, and needs of the SAF's cadre members can be addressed, which in turn contributes to their job satisfaction. As increased job satisfaction is vital to staff retention in the SAF, the connection between communication and retention has been established. Whether there is also a link between the recruitment of cadre members and communication is difficult to say, because the SAF are based on a militia system and conscription and thus do not recruit in the actual sense. However, it is fair to say that good communication will foster rather than hinder members of the SAF to join their cadre personnel.

It goes without saying that the 34 interviews, which are the empirical foundation of this qualitative research, are by no means representative of the whole of the SAF cadre personnel. Neither is the category system which served as the base for analysing the interviews conclusive. It merely mirrors the statements of those members of the SAF's cadre personnel who were interviewed, based on the researchers' understanding and explanation. As is always the case in qualitative research, the role of the researchers is to keep as neutral as possible and not to include personal interpretation patterns. Thus, by developing the category system of this study in pairs, we hope that we were able to minimise a researcher's effect.

References

Baroncelli L, Freitas A (2011) The visibility of the self on the web: a struggle for recognition. In: Conference Paper for ACM WebSci '11, 3rd International Conference on Web Science, Koblenz, Germany, June 14-17, 2011 [online]. [Viewed 2 December 2019]. Available from http://www.websci11.org/fileadmin/websci/Posters/191_paper.pdf

Bartels J, Peters O, de Jong M, Pruyn A, van der Molen M (2010) Horizontal and vertical communication as determinants of professional and organisational identification. Pers Rev 39(2):210–226

Böhm A (2007) Theoretisches Codieren in der Grounded Theory. In: Flick U, von Kardorff E, Steinke I (eds) Qualitative Forschung. Ein Handbuch, 5th edn. Rowohlt, Reinbek b. Hamburg, pp 475–485

Boons M, Stam D, Barkema HG (2015) Feelings of pride and respect as drivers of ongoing member activity on crowdsourcing platforms. J Manag Stud 52(6):717–741

Camilleri E, Van der Heijden BIJM (2007) Organizational commitment, public service motivation, and performance within the public sector. Public Perform Manag Rev 31(2):241–274

Chan SHJ, Lai HYI (2017) Understanding the link between communication satisfaction, perceived justice and organizational citizenship behavior. J Bus Res 70(2017):214–223

Devereux L, Melewar TC, Foroudi P (2017) Corporate identity and social media: existence and extension of the organization. Int Stud Manag Organ 47(2):110–134

Fahmy S, Cho S, Wanta W, Song Y (2006) Visual agenda-setting after 9/11: individuals' emotions, image recall, and concern with terrorism. Vis Commun Q 13(1):4–15

Federal Constitution of the Swiss Confederation of 18 April 1999 (1999; Status as of 1 January, 2020). [Viewed 14 January 2020]. Available from https://www.admin.ch/opc/en/classified-compilation/19995395/index.html#ani1

Felfe J, Schmook R, Six B (2006) Die Bedeutung kultureller Wertorientierungen für das Commitment gegenüber der Organisation, dem Vorgesetzten, der Arbeitsgruppe und der eigenen Karriere. Z Pers 5(3):94–107

Flick U (2004) Qualitative Sozialforschung. Eine Einführung, 4th edn. Rowohlt, Reinbeck b. Hamburg

Gautam T, Van Dick R, Wagner U (2004) Organizational identification and organizational commitment: distinct aspects of two related concepts. Asian J Soc Psychol 7:301–315
Gioia DA, Corley KG, Hamilton AL (2012) Seeking qualitative rigor in inductive research: notes on the Gioia methodology. Organ Res Methods 16(1):15–31
Grimmelikhuijsen S (2012) Linking transparency, knowledge and citizen trust in government: an experiment. Int Rev Adm Sci 78(1):50–73
He H, Brown AD (2013) Organizational identity and organizational identification: a review of the literature and suggestions for future research. Group Org Manag 38(1):3–35
Ho AT-K, Cho W (2016) Government communication effectiveness and satisfaction with police performance. A large-scale study. Public Admin Rev 2017 77(2):228–239
Jones C, Volpe EH (2010) Organizational identification: extending our understanding of social identities through social networks. J Organ Behav 32(3):413–434
Kandlousi NSAE, Ali AJ, Abdollahi A (2010) Organizational citizenship behavior in concern of communication satisfaction: the role of the formal and informal communication. Int J Bus Manag 5(10):51–61
Karanges E, Johnston K, Beatson A, Lings I (2015) The influence of internal communication on employee engagement: a pilot study. Public Relat Rev 41(1):129–131
Karnica T, Kumar A (2019) Employer brand, person-organisation fit and employer of choice. Pers Rev 48(3):799–823
Kim T, Chang K, Ko YJ (2010) Determinants of organisational identification and supportive intentions. J Mark Manag 26(5–6):413–427
Kissel P, Büttgen M (2015) Using social media to communicate employer brand identity: the impact on corporate image and employer attractiveness. J Brand Manag 22(9):755–777
Mayring P (1994) Qualitative Inhaltsanalyse. In: Böhm A, Mengel A, Muhr T, Gesellschaft für Angewandte Informationswissenschaften (GAIK) e. V (eds) Texte verstehen: Konzepte, Methoden, Werkzeuge. UVK (Schriften zur Informationswissenschaft 14), Konstanz, pp 159–175
Mayring P (2000) Qualitative Inhaltsanalyse. Forum qualitative Sozialforschung [online]. [Viewed 16 December 2018]. Available from http://qualitative-research.net/fqs/fqs-d/2-00inhalt-d.htm
Mercurio ZA (2015) Affective commitment as a core essence of organizational commitment: an integrative literature review. Hum Resourc Dev Rev 14(4):389–414
Mujib H (2017) Organizational identity: an ambiguous concept in practical terms. Admin Sci 7(3):28 [online]. [Viewed 29 October 2019]. Available from: https://www.mdpi.com/2076-3387/7/3/28
Pleil T, Matthias B (2017) Soziale Medien in der externen Organisationskommunikation. In: Schmidt J-H, Taddicken M (eds) Handbuch Soziale Medien. Wiesbaden, Springer Fachmedien, pp 129–149
Rager G, Oestmann I, Werner P, Schreier M, Groeben N (1999) Leitfadeninterview und Inhaltsanalyse. In: Viehoff R, Rusch G, Segers RT (eds) SPIEL: Siegener Periodicum zur Internationalen Empirischen Literaturwissenschaft, 18 (1). Peter Lang, Frankfurt a.M., Berlin, Bern, Bruxelles, New York, Wien, pp 35–54
Scott CR (2007) Communication and social identity theory: existing and potential connections in organizational identification research. Commun Stud 58(2):123–138
Shamir B, Kark R (2004) A single-item graphic scale for the measurement of organizational identification. J Occup Organ Psychol 77:115–123
Stocker D, Jacobshagen N, Semmer NK, Annen H (2010) Appreciation at work in the Swiss Armed Forces. Swiss J Psychol 69(2):117–124
Strauss A (2004) Methodologische Grundlagen der Grounded Theory. In: Strübing J, Schnettler B (eds) Methodologie interpretativer Sozialforschung. Klassische Grundlagentexte. UVK, Konstanz, pp 429–451
Suchman L (1995) Making work visible. Commun ACM 38(9):56–63
Szvircsev Tresch T, Wenger A, De Rosa S, Ferst T, Giovanoli M, Moehlecke de Baseggio E, Reiss T, Rinaldo A, Schneider O, Scurrell JV (2019) Sicherheit 2019: Aussen-, Sicherheits- und

Verteidigungspolitische Meinungsbildung im Trend. ETH Zurich/Military Academy at ETH Zurich, Zurich
Treem JW, Leonardi PM (2013) Social media use in organizations: exploring the affordances of visibility, editability, persistence, and association. Ann Int Commun Assoc 36(1):143–189
Wiesenfeld BM, Raghuram S, Garud R (1999) Communication patterns as determinants of organizational identification in a virtual organization. Organ Sci 10(6):777–790
Zerfass A, Verhoeven P, Moreno A, Tench R, Vercic D (2016) European communication monitor: exploring trends in big data, stakeholder engagement and strategic communication. Results of a survey in 43 countries. EACD/EUPRERA, Quadriga Media Berlin, Brussels

Part II
Gender-Specific Representation on Social Media

Managing Femininity Through Visual Embodiment: The Portrayal of Women on the Instagram Accounts of the Swedish and the Swiss Armed Forces

Andrea Rinaldo and Arita Holmberg

Abstract As a gendered organisation, the military's organisational identity is based, among other things, on what is considered as stereotypic masculinity: strong, brave, and tough men represent the ideal warrior. The increasing number of female soldiers threatens this part of the organisational identity. Social media such as Instagram serves as a means to reflect organisational identity. Therefore, the following contribution examines how the armed forces deal with the gendered nature of the military by comparing the portrayal of women on the Instagram profiles of the Swedish and the Swiss Armed Forces. In consideration of the societal context as well as both countries' military characteristics, we analyse whether gender stereotypes in the military are highlighted or reduced and how the female body is treated in relation to military identity. A visual content analysis of the Instagram posts from 2018 reveals that women are portrayed in stereotypical masculine roles in the case of the Swedish Armed Forces, whereas the Instagram pictures of the Swiss Armed Forces portraying women highlight stereotypical female attributes.

Keywords Gender embodiment · Visual content analysis · Social media · Military · Organisational identity

A. Rinaldo (✉)
Military Academy (MILAC) at ETH Zurich, Birmensdorf, Switzerland
e-mail: rinaldoandrea@gmx.ch

A. Holmberg
Department of Security, Strategy and Leadership, Swedish Defense University, Stockholm, Sweden
e-mail: arita.holmberg@fhs.se

E. Moehlecke de Baseggio et al. (eds.), *Social Media and the Armed Forces*, Advanced Sciences and Technologies for Security Applications,
https://doi.org/10.1007/978-3-030-47511-6_5

1 Introduction

Social media offers organisations new means to promote their identity among their target audience and to convey their messages through images. Organisational use of social media can affect the way organisations are perceived externally, for instance by the population. It can also spur internal dynamics and discussion. Research on social media and military organisations has so far focused on the role of social media as a strategic narrative and a channel for public communication. At the individual level, soldiers' experiences of war have also been analysed. However, there is less research on social media as an instrument or reflection of organisational identity.

Historically and politically linked to civil rights and duties, the military is a gendered organisation in which traditional images of masculinity dominate. The current transformation of the armed forces and their tasks – which require more and more non-traditional combat skills – as well as the increasing admission of women to the military in general and to combat functions in particular challenges this hegemonic masculinity. It is likely that social media plays a role in this regard as a reflection of organisational identity: either by (re-)producing or by breaking gender stereotypes in the military. However, this needs to be explored further.

The objective of this contribution is to compare the portrayal of women on the Instagram accounts of the Swedish and the Swiss Armed Forces in consideration of the societal context as well as both countries' military characteristics, with special focus on the inclusion and participation of women. The armed forces of Sweden and Switzerland are alike in many respects. They have similar goals and tasks – such as peacebuilding and subsidiary support of civilian authorities in addition to national defence – and both have conscription systems. However, when reintroducing conscription in 2018 after having tried out a professional system since 2010, Sweden chose to make its military gender-neutral. Furthermore, both men and women can still join the armed forces voluntarily in Sweden. Thus, we are comparing two countries one of which aims to include women, whereas the other conscripts only men while keeping military service for women voluntary. As state authorities, both countries' armed forces are, of course, bound by their respective equality laws and regulations. At this point, it is worth mentioning that women's suffrage was introduced in 1921 in Sweden and only 50 years later, in 1971, in Switzerland. Moreover, same sex marriage in Sweden has been possible since 2009, but is not allowed in Switzerland, which only offers registered partnership for same-sex couples. These differences in the development of gender equality might affect the position of women (and other non-dominant groups) in the military. Even though both countries aim to increase the proportion of women in the military organisation, according to Bondolfi (2012), conscription that only applies to men leads to an imbalance of the sexes, especially in upper cadre positions. Furthermore, without equal regulations for men and women concerning compulsory military service, actual equality between the sexes in the armed forces is difficult to achieve (Bondolfi 2012). In addition, with the temporary suspension of the compulsory military service, the

Swedish military organisation was forced to adapt to the recruiting conditions of other public and private organisations, and to be able to attract both sexes. Such differences can affect the identities of the different military organisations. This raises the question of how women are portrayed on the official Instagram account of the Swedish and the Swiss Armed Forces, and whether the difference in the recruitment of women is recognisably transmitted to this online communication platform. We expect the two countries' different backgrounds regarding gender policies and equal opportunity as well as the different recruiting strategies to be reflected in the way the respective armed forces portray women on Instagram.

We consider women's portrayal on the official Instagram accounts of armed forces an important element for the research on gender equality in the military as well as for armed forces communication research. Firstly, the significance of the portrayal is reinforced by the fact that images make up the largest part of Instagram by far, whereas texts are of little importance. Secondly, image contents are shaped by narrative organisational patterns and affect the recipient's perception of the portrayed content. The Instagram posts were analysed using a quantitative as well as a qualitative approach for visual analysis. The discussion of the results addresses the different ways in which gendered portrayals are used by military organisations, either by reproducing or breaking gender stereotypes.

The present contribution is structured as follows: firstly, we introduce previous research and theories on the military as a gendered organisation. Here, the centrality of the military body becomes apparent. In addition, research on social media and the military is discussed. In the next section, the method and material of the study are presented. Thirdly, we present the results of each country separately before a comparison is made. Finally, conclusions are drawn and suggestions for further research presented.

2 Theoretical Frame

2.1 The Military as a Gendered Organisation

Armed forces all over the world are considered gendered organisations (Addelston and Stirratt 1996; Sasson-Levy 2011; Muhr and Slok-Andersen 2017; Alvinius and Holmberg 2019). A major trait of military identity is masculinisation (Hearn 2011; Hale 2012). Masculine or gendered identities are not individual characteristics, but are constructed in social interaction (Goffman 1959) and can be reinforced and reproduced by institutions (Hinjosa 2010; Acker 1990). It could therefore be assumed that feminisation is not a common feature in military organisations. Although the image of the strong, brave, tough warrior does not apply to every soldier, this stereotype of the ideal and hegemonic warrior nevertheless prevails (Duncanson 2009). As women are now allowed to join the military, this part of the organisational identity is being threatened (Sasson-Levy 2011), as is also noted by Pin-Fat and Stern:

> Gendered divisions, because of their associations with the "natural," provide a powerful mechanism for creating seemingly stable categories, or zones of distinction. Gender coding creates a "natural" order of distinctions whose grammar serves as an organizing principle for political life. Attempts at sustaining boundaries between the military and civilian life, men and women, war and peace, and so on reveal how these boundaries rely on clear coding of masculinity and femininity and how the taken-for-granted identity of military and the boundaries upon which it rests are unsettled by the inclusion of the feminine (Pin-Fat and Stern 2005: 34).

Pin-Fat and Stern (2005) argue that a gendered distinction is necessary for upholding the myth of the military as protecting the nation and soldiers sacrificing their lives for this cause. The feminisation of the military (by introducing female soldiers) challenges this idea in fundamental ways. Therefore, it is particularly interesting to study how women are embodied in the military, and how the female body is dealt with in relation to military identity.

The construction of the military body as masculine has been extensively researched. For instance, allowing young people to enter into uniform has been interpreted as a technique used by government to produce subjectivities in them by performing masculinity (Wells 2014). It appears that a woman in the military either has to work to become a man and practice hyper- masculinity (Höpfl 2003; Sasson-Levy 2011) or be assigned a female role. Here the rescue of private Jessica Lynch may serve as an example. Pin-Fat and Stern outline this media-effective, staged story, discussing how it reinforces and unsettles the masculine identity of the military by representing "Lynch as a woman in the fighting ranks" (2005: 27). Previous research thus emphasises that soldiers are embodied in a context of what we perceive as either masculinity or femininity. The question is whether there is a possibility of breaking this reproduction of gender stereotypes, and if so, what this would look like.

Welland (2017) argues that the (hyper-)visibility of military bodies is important in the context of today's military conflicts and serves as a purpose to legitimise them. She conceptualises the military body in this context as a *liberal warrior*, pictured as being capable of performing a wide range of tasks. The liberal warrior has been heavily exposed and embodied as both vulnerable and violent/powerful in relation to the conflicts in the Middle East – while civilian, distant bodies are rendered invisible (Welland 2017). The masculine military body is also part of the popular image of the military, supported by the vast representations of the male military body in film and literature (Godfrey et al. 2012).

McSorley (2013a, b) identifies three dimensions of the war/body nexus: preparing for war, practices of war and the aftermath of war, with an emphasis on the fitness of the military body. Even though this emphasis is widely recognised in the literature (Newlands 2013; Burridge and McSorley 2013), due to new technology and changes in the way war is conducted, the military body has somewhat lost its need for physical strength (Godfrey et al. 2012; McSorley 2013a, b). The traditional idea of the fit soldier is likely to be upheld by the military organisations themselves as a form of resistance towards an inclusive military that allows for different bodies, which risk fragmenting the organising principle of the masculine soldier, thereby affecting the military organisational identity.

2.2 *Social Media and the Military Organisation*

With the introduction of new media, social media communication has become an important channel for armed forces to convey their organisational identity (Andrén-Papadopoulos 2009; Hellman and Wagnsson 2013; Hellman 2016; Hellman et al. 2016; Lawson 2014). Social media offers new opportunities for the armed forces "to communicate and interact directly with the population" and gain legitimacy (Moehlecke et al. 2019: 44). At the same time, social media communication takes place in a context that is surrounded by more societal and political pressures as to how the military should behave, for instance regarding gender equality (Holmberg and Alvinius 2019).

Recent research shows that the Swedish military represents Sweden's identity as a gender-exceptional nation in its marketing campaigns. In this context, Swedish sexuality rights are pictured as something the military defends – thus contributing to building legitimacy and relevance for the military (Strand and Kehl 2018). Since the central focus of the social media user's interest is on pictures, this leads to new means of virtual communication and interaction (Autenrieth 2014; Traue 2013). Therefore, it is interesting to analyse how the gendered nature of the military organisation is expressed through a visual narrative – in particular, how women and their bodies are portrayed, and whether gender stereotypes are highlighted or reduced, because the media generally tends to portray men and women in a way that fosters gender stereotypes (Wood 1994).

A study of how Swedish women working in the military portray themselves on their personal Instagram accounts finds that women use social media to present alternative images of themselves (Lundqvist 2018). Lundqvist (2018) argues that these women use the objectification of themselves to challenge stereotypes as well as the idea of who can be part of the military. Instagram thus offers an arena to empower these women. They refuse to conform to the masculine identity by highlighting their femininity – often through make up (Lundqvist 2018). This can be interpreted as a form of resistance against the masculine identity of the military organisation. In our study, instead of analysing the self-portrayal of female armed forces members, we pursue the question of how women are portrayed *by* the armed forces, and thus how the latter use this arena with regard to communicating their organisational identity.

3 Data and Method

As was noted in the theoretical section above, bodies are central for military organisations. Feminist security studies have recognised this, and a 'corporeal turn' is also appearing in general international relations. The body is studied by means of visual discourse analysis that can focus on either affect, emotions or the somatic (Mutlu 2013). In the present study, the body is central, as its representation is seen as expressions of military identity.

In order to answer our research question, we analysed the pictures approved by the Swedish and the Swiss military that were published in 2018 on the social media platform Instagram (Sweden: n = 262; Switzerland: n = 576). We applied a visual content analysis to study the portrayal of women, following a standardised procedure to structure the material by categorising the images based on predefined categories. We distinguished between pictures showing either male or female persons or both (if recognisable) or no persons at all. According to Rössler (2010), quantifying and descriptively analysing certain motives or people in pictures is one of five epistemological concerns of visual content analysis. In this study, the aim was to single out pictures containing persons identified as women.

We are aware that this method of data collection contains a certain element of reproducing stereotypic norms about how women and men usually appear. For instance, a long-haired, small-figured person with their back to the camera was categorised as a woman. This is problematic. Moreover, it is also problematic to categorise individuals as women although they may not necessarily identify as such. However, this method of data collection was considered to be the only one available. In fact, this approach is not unusual, seeing as the representation and stereotyping of gender or ethnical minorities is one of the most common topics of visual communication research studies applying visual content analysis. Frequency analyses in this context are often linked to and compared with the actual representation of the investigated population, the portrayed persons being analysed within their (often stereotypical) context of action (Lobinger 2012). We also recognise that we as researchers are part of the gendered discourse and thus cannot stand above it. For the analysis of the pictures, we refer to gender stereotypes as they appear in the literature. As was mentioned in the previous section, the image of an ideal soldier is associated with traditionally and stereotypically masculine characteristics (Duncanson 2009). This means that we define femininity with reference to how masculinity is defined in the military and as complementary to the latter. This way, we may attempt to discuss the stereotypic gender discourse instead of merely accepting it.

Since we are interested in the portrayal of gender, only images portraying no more than two persons in focus and containing at least one woman were further analysed (Sweden: n = 45; Switzerland: n = 28). This decision was based on the condition that the gender of the portrayed persons must be identifiable and on the view that, in order to discuss gender portrayal, persons as individuals need to be the main focus of the image. Our main country samples thus exclude all pictures showing more than two persons or no persons at all, as well as pictures showing one or two persons with an unassignable gender, for example due to the photographic angle or to bodies and faces being covered by helmets, protective suits or similar.

In addition to analysing gender distribution in general, we used inductively generated categories for a more specific frequency analysis of gender on the respective Instagram accounts. Looking at all pictures in the Swedish and the Swiss samples, we identified different clothing (military clothing (i.e. camouflage suits and other uniforms), civilian clothing, sportswear, and specific work clothing), different facial expressions (presence or absence of smile), weapons and weapons systems (presence or absence of weapons), animals (presence or absence of animals), and different scopes of activity. For the categorisation of the latter, we referred to recognisable surroundings and contextual factors, but not to the accompanying text of the images. This means that the assigned activities do not necessarily correspond to the actual activity executed in the moment the photo was shot.

We used standardised visual content analysis as a basis for further, qualitative analyses, pursuing an explorative approach. This second step allowed us to go beyond the quantification of image motives and address the performative power that organisations have when they communicate via social media. In the case of Instagram, reality is constituted by visual discourses, which affect social and institutional structures. We therefore applied visual discourse analysis to focus on the embodiment of gender and mediated gendered identities on the posted images of the Swedish and the Swiss Armed Forces Instagram accounts. In accordance with our research question, we only focused on pictures showing at least one woman in the qualitative part of our analysis.

Even though images on Instagram – like most other visual media contents – are accompanied by text, most studies using visual content analysis refrain from taking into account this multimodal context (Lobinger 2012). Hence, we decided to focus only on the visual content, because images take up by far the largest part of Instagram, whereas verbal elements are of little importance, thus reinforcing the visual effect.

4 The Swedish Case

4.1 The Swedish Context

While the Swedish Armed Forces focused mostly on international operations during the 2000s, they have been re-oriented towards the task of territorial defence since around 2015. In an effort to professionalise the military, conscription was abandoned in 2010 (Holmberg 2015). However, voluntary recruitment proved difficult, and conscription was re-introduced in 2018 (voluntary participation remains an option). This time conscription was made gender-neutral. The number of female conscripts amounted to 15.5% in 2018 (about 3700 conscripts started military education that year). Among military personnel, 7% of the officers are women, and of the civilians, 40%. In total, this amounts to 18% women among the continuously serving personnel (Swedish Armed Forces 2019). While the Swedish Armed Forces only published their first picture on Instagram in 2013, a social media policy was adopted as far back as 2011, identifying both opportunities and risks. However, the advantages dominate – such as increasing the public's knowledge of the Swedish Armed Forces and contributing to an open discussion on their role and tasks in society (Swedish Armed Forces 2011).

4.2 Data Collection and Analysis

Data was collected by going through all the posts in the Swedish Armed Forces Instagram account from 1 January to 31 December 2018. The pictures containing persons identified as women were singled out and collected in a word file, along with the date of publication and the accompanying text. In addition, pictures that have been 're-grammed' by the Swedish Armed Forces from other Instagram accounts were included, because the Swedish Armed Forces chose them to represent their main account, although they originated elsewhere.

The total number of pictures published on the official Instagram account of the Swedish Armed Forces during 2018 was 262.[1] The total number of pictures portraying women was 103, which makes 40%. Already at this stage, we can thus conclude that pictures of women in social media constitute a larger share than the actual share of women within the organisation (18% in total, 7% in military personnel). As was mentioned above, the sample for the further analysis includes only pictures showing a maximum of two persons and portraying at least one woman. In the process of coding, each picture in the data file was systematically analysed and coded accordingly. One picture was eliminated because it pictured two people from behind, at a distance, and was thus too difficult to analyse. This left us with a sample of 45 pictures portraying at least one woman in focus. The results were then summarised in an excel file, and an overall impression of the quantitative result was gathered, which is presented in the analysis below. However, a more qualitative analysis was also pursued by adding a cultural connotation to the pictures, as was a visual discourse analysis focusing on gender and embodiment.

4.3 Results

Women and men are portrayed together in around 40% of the posts studied. A majority of the portrayal of women is thus photographs of individual women. Most women in our sample are portrayed in camouflage clothing (67%) or other military uniforms (17%). Only a minority of 8% is shown in other clothes (see Table 2 in the comparison section). In some pictures of women, clothing is not recognisable (8%). With regard to facial expressions, women are portrayed both smiling and very sternly, however, images without smiles clearly dominate. There are not many posts with animals accompanying women, only a few photos of dogs. Weapons and weapons systems are present in a majority of the pictures of women posted on the Swedish Armed Forces Instagram account.

Below are three pictures which were selected in order to illustrate the typical image of a woman on the Swedish Armed Forces Instagram account. The analysis is expanded in relation to each picture.

[1] https://www.instagram.com/forsvarsmakten/?hl=sv

Fig. 1 Posted on 29 October 2018. (Photographer Rickard Törnhjelm/Swedish Armed Forces)

In Fig. 1, the woman is portrayed as focused and in a passive position – on guard with a stern face, although a repressed smile can be detected. She has a weapon (rifle) to accompany her. This makes us associate the image with an active context: she has a role to fulfil. The woman is embodied as a prepared soldier on alert. The picture is an example of the many Instagram posts in the Swedish sample that show women with weapons (see Table 1 in the comparison section).

The setting in Picture 1 is a forest, a traditional territorial defence setting. In the Swedish sample, women are most often portrayed in uniform, either in camouflage or naval uniform. The example above depicts a passive position, but there are also many examples of women taking active positions. It is clear that the focus of the visual portrayal on the Swedish Instagram is on the somatic.

A clear majority of the analysed photographs are taken outside in 'active environments', such as the forest or at sea in a geographic setting that looks like Sweden. There are both summer and winter environment pictures. This could be seen as a way of transferring an image of the main trait of the current organisational identity of the Swedish Armed Forces: territorial defence as the main task– a transformation process that has been ongoing since around 2015.

Women did not prove to be overly associated with animals in the sample (see Table 1 in the comparison section). There are a few pictures of women and dogs, such as the one below (Fig. 2). In this picture, the woman and her dog are portrayed in full gear, most likely prepared for some mission or exercise. There are also images of women and men with horses, but these have been left out of the sample due to the presence of multiple persons. However, a small note on the relationship between horses and gender may be in order here. In Sweden, horse-riding is a popular sport dominated by women and girls – who since the 1970s have come to take over the

Fig. 2 Posted 6 June 2018 (Photographer Bezav Mahmod/Swedish Armed Forces)

traditional association between horses, men, and the military. It is therefore likely that the images of horses and women in a military context are used to attract women to the military. The Swedish Armed Forces have in recent years conducted recruitment campaigns at major events such as the Stockholm International Horse Show (show-jumping and dressage), and present themselves and their cavalry at the award ceremonies of many major competitions in different branches of the sport.

In some pictures, women wear civilian clothes, for instance when portrayed as new recruits or at sports events. These pictures embody women as capable military bodies, able to perform physically or to carry heavy weights. In these cases, women are sometimes smiling, which shows that an emotional aspect is allowed. However, this is tightly controlled. For instance, smiles are interpreted as joy and anticipation in the face of military service. Happiness is mostly expressed in the context of social gatherings, where the persons in the picture are portrayed talking and laughing. Here, the posts can be interpreted as symbolising connection within the group. It is attractive to be part of a group and to have a nice time together, to belong by means of similar clothing and to focus on a mission. Both women and men are allowed to

Fig. 3 Posted on 9 November 2018 (Photographer Alexander Gustavsson/Swedish Armed Forces)

symbolise this group connectivity. However, the majority of pictures contains images of women categorised as showing 'no smile' (see Table 1 in the comparison section).

Most women are young and would be perceived by many as good-looking, but there are also some pictures of older women. When older women are portrayed, they are most often interpreted as civilians (relatives or other civilian employees or volunteers). Women who are portrayed individually on the sample pictures very often carry symbols of the military organisation's status and power: they are in full uniform, carrying pack, camouflage, rifle; they are standing next to tanks (in some cases accompanied by men), naval vessels or fighter jets, as shown in Fig. 3. They are portrayed both from the front and from behind. In many of the pictures, women are being visually associated with power, strength, and endurance. As a result, the female body is placed alongside the male body without highlighting the female gender. What we see is a 'war machine'. Overall, picturing women this way can be seen as a way of disciplining them, making them submit to masculinity. That way, they become acceptable to the organisation through a form of de-feminisation. Thus, the Swedish case is in line with the results presented by Höpfl (2003) and Sasson-Levy (2011).

It is sometimes difficult to interpret facial expressions, which could be connected to a culture of submission of emotional expressions. Often, very neutral facial expressions are portrayed, an observation which can also be referred to the masculine identity of military organisations.

The symbols, colours, weapons and structures of the military organisation are embodied in the photographs through the participation of women. Women are also present in public relations campaigns – for example, regarding value-based themes such as pride/LBGTQ+ – as well as in recruitment campaigns. As was noted in previous research, this is an image that the Swedish Armed Forces pursue as a reflection of the national image of Sweden (Strand and Kehl 2018). In addition, when it comes to territorial defence, women are present; in fact, they are often pictured alone. It is probably a conscious strategy to include both men and women on the official Instagram account and to try to picture them in similar ways. In reality, however, the number of women in the Swedish Armed Forces is much lower than might be deduced from the Instagram account.

It was noted that the majority of the posted pictures contain young people in their twenties. This signals the need of the military organisation to portray young, healthy individuals – the type of military body that it wishes to recruit. Older persons are clearly not the target demographic of this Instagram account.

5 The Swiss Case

5.1 The Swiss Context

Even though the military service in Switzerland is compulsory for men, the Swiss Armed Forces are struggling with recruitment – and with refraining recruits from leaving the armed forces for the alternative civil service, which is considered more attractive. In order to fight decreasing application numbers, the Swiss Armed Forces launched an Instagram account in 2017 with the aim of attracting the attention of the younger generation. Its main target group of followers is young people and explicitly includes both men and women. As is the case in other departments of the Federal administration of Switzerland, the armed forces seek to increase the share of women. Today, the military counts 0.8% female members. This number is extremely low compared to the share of women in other European armed forces, even when taking into account that in Switzerland – contrary to Sweden – conscription applies only to men. With the introduction of the reform *Armee 95* in 1995, women were fully integrated into the Swiss Armed Forces. They obtained the same rights and duties as men, with the exception of admission to combat troops, which was only granted in 2004. Although today military service is still optional for women, they have to pass the same assessments as men. In return, they receive the same education and are now admitted to all combat troops and can achieve all military ranks (Paladino 2015; Seewer 2003). However, political and public discussions reveal that the equal integration of women in the armed forces is associated with worries, prejudices, and resistance. For instance, a representative survey of Swiss citizens from 2015 showed that only 30% of the population agree on the expansion of compulsory military service to women (Szvircsev Tresch et al. 2015). Furthermore, in another representative survey Szvircsev Tresch et al. (2019) found that 55% of Swiss citizens consider women physically unsuitable for certain tasks in the armed forces. This raises the question if and how the imbalance in recruitment, the gender distribution in the military, and the concerns regarding women's physical ability for military service are reflected on the Instagram account of the Swiss Armed Forces.

5.2 *Data Collection and Analysis*

The data basis for the Swiss case consists of all Instagram posts published in 2018 on the official account of the Swiss Armed Forces.[2] All posts with the accompanying text and date of publication were collected in MAXQDA,[3] a software for qualitative, quantitative, and mixed methods data analysis. The total number of pictures published between 1 January and 31 December 2018 is 576. 28 pictures show a maximum of two persons and at least one woman, constituting the Swiss sample for further analysis. In total, there are 219 pictures portraying a maximum of two persons in focus. Even though women occur much less often on the Instagram account of the Swiss Armed Forces than men, they are still overrepresented considering the actual number of women serving in the Swiss Armed Forces today (0.8%; Swiss Armed Forces 2019). The vast majority of the pictures show young women and men in their twenties, which probably corresponds to the organisation's intended target audience.

5.3 *Results*

Half of the 28 pictures show at least one woman in camouflage clothing, whereas 13 pictures (46%) portray women wearing civilian clothing, sportswear or work clothing that is sometimes also recognisable as a specific military uniform to the trained eye. In two of the pictures (7%), clothing is not recognisable (see Table 2 in the comparison section). The majority of women portrayed on the Instagram account of the Swiss Armed Forces in 2018 is wearing camouflage and thus a type of clothing that is unique to the armed forces and a distinguishing feature of the military organisation.

The contents and activities depicted in the data sample were inductively coded and grouped into categories. We found women in a hospital or *caring* environment, women working or posing with *animals*, women portrayed as *top-class athletes*, and women in a conversation situation, usually *informing* other women about something. Most pictures were assigned to these four categories. We also found that women are always pictured smiling on posed photographs in which they look directly into the camera. On most snapshots taken while performing an activity, they focus on their work and seem to be concentrated, and a smile is not present. The presence or absence of a smile therefore seems to be connected to the context of the activity captured in the picture. In total, we found 12 (43%) women smiling on 28 pictures.[4] This finding indicates that in the pictures of the Swiss Armed Forces'

---

[2] https://www.instagram.com/armee.ch/

[3] https://www.maxqda.de/

[4] In this regard, we wish to point out the findings based on our extended sample of the Swiss case, which also includes pictures with male subjects. Out of 167 pictures, we found 32 showing a man with a smile. The share of smiling men (19%) is therefore much lower than that of smiling women (43%).

Instagram account, women appear to be rather passive, friendly looking, and engaged in rather calm activities. This observation is supported by the fact that 12 pictures of our sample were taken outdoors, whereas 16 show offices or classrooms, sports halls, hospital environments or other indoor locations that are hard to identify. Pictures taken outside show rural environments, however, most often the surroundings are indistinct or barely recognisable in the image background.

Even though images portraying only men were not included for further analysis in this study after the initial categorisation, even cursory analysis reveals more elements indicating a specific military context in those pictures – for instance, through the presence of tanks, rifles or military training facilities in the image background – than in the pictures portraying women. In this regard, the absence of weapons in our sample is striking. While pictures of men holding a rifle or a gun are prominent, no picture portraying women in the same manner was found. In fact, the only picture of 2018 portraying a woman holding a weapon is the following:

In Fig. 4, the woman is clearly identifiable as a member of the Swiss Armed Forces based on her camouflage suit. The scene of the picture could be a march; the identification number on her comrade in the background might be indicating an

Fig. 4 Posted on 18 September 2018

orienteering. The woman is walking along a street in a rural environment. In the background, there are trees and mountains, and vehicles do not seem to pass frequently, which allows her to walk in the middle of the street. This photo is one of the few in the Swiss sample where a woman is portrayed in an outdoor environment with a recognisable background. She is accompanied by a goat, which – in combination with her smile – adds a certain cute factor to the picture. As was mentioned above, this is one of the pictures where the woman is smiling while looking directly into the camera. Based on her expression, she seems to be enjoying the assigned task, thus conveying a positive message. She seems to be holding her rifle in a casual fashion, and her hat is not placed properly on her head. Therefore, the scene pictured here seems to lack seriousness, which is rather inappropriate for the military. On the other hand, these elements manage to bridge the civil/military gap to a certain degree by showing that the borders between the soldier and the civilian are fluid.

Animals are found on three further images of our sample, one showing a woman with a service dog, a shepherd (Fig. 5), and two showing women caring for a husky and its puppy (Fig. 6).

Figures 5 and 6 have just as many common features as they do differences. Both pictures portray a young woman crouching on one knee next to a dog with one hand around the dog's neck, thereby showing affection for the animal. Both photographs were taken outside.

The woman in Fig. 5 is wearing a camouflage suit, which clearly identifies her as a member of the armed forces. The dog next to her indicates that she is probably a

Fig. 5 Posted on 27 August 2018 (Photographer Andrea Soltermann)

Fig. 6 Posted on 5 September 2018

dog handler, which is a popular unit among women who join the military. She is looking sideways, away from the dog, resting one arm on her knee, and is smiling slightly. This posture gives the impression of a mixture of general coolness and control over the shepherd dog, which in turn seems to empower her as a soldier.

Despite the commonalities of the pictures, the scenes show a different content. The woman in Fig. 6 is looking into the camera, smiling more obviously. The husky and its puppy most certainly have a different effect on the viewer then the shepherd dog. Furthermore, we find no reference to the military organisation in Picture 6. The portrayed woman is wearing a specific work uniform, which, in combination with the image background, as well as the dogs she is petting, suggests that she might be working in a shelter caring for animals or a dog training facility.

As was already mentioned, one activity that was identified on various images in the sample is the exchange or provision of information, as portrayed in Fig. 7. The posters in the background suggest that the scene is an information event of education institutions, at which the Swiss Armed Forces also seem to distribute information about themselves. We find various pictures in the Swiss sample portraying women explaining or showing something to other women.[5] Selecting women for this task could be based on the aim to make the armed forces more attractive to women. However, it might also convey the impression that women and men are treated differently, although women formally are fully integrated into the Swiss Armed Forces. Furthermore, it raises the questions whether only women are suited

[5] Usually, only one person is shown in such pictures, unlike in Fig. 7.

Fig. 7 Posted on 1 September 2018

to introduce the military to other women, and whether women are also instructed to inform men, and vice versa.

Other pictures show women in a hospital environment. Many women in the Swiss Armed Forces chose to join or are assigned to medical services in the military. In fact, in Swiss society in general, paid and unpaid care work is most often carried out by women (Eidgenössisches Büro für die Gleichstellung von Frau und Mann EBG 2010). In these pictures, there are usually hardly any indications that the women portrayed are members of the armed forces, except for specific clothing in a few cases. Portraying female members of the armed forces in a profession or task typically associated with women can be seen as a form of highlighting femininity, which fosters gender stereotypes, especially when the scenes are also accentuated by the women's friendly smiles and the interaction with the camera, as found in Figs. 4 and 6. There are also images of women showing clear association to the military by portraying them in their military uniform. However, there are no other elements found in most of these pictures that refer to the military organisation.

6 Comparison and Discussion

This study shows, unsurprisingly, that social media has become a fully-fledged tool for military organisations in conveying their organisational identities. As such, social media has become an important source of inquiry for research on the military organisation.

The Instagram analysis of the Swedish and the Swiss Armed Forces showed that on both accounts, women are overrepresented compared to the actual number of serving women. In the Swedish sample, women appear on 40% of the pictures showing one or two people in focus, while the number of women in the Swedish Armed Forces is 18% (including civilians). On the Swiss Armed Forces Instagram account, the share of women is 13% (which is 16 times higher than the share of women in the Swiss Armed Forces). Beside this commonality, the findings of the case studies show two rather different portrayals of women.

Referring to McSorley's (2013a, b) dimensions of how bodies are presented in the context of war (preparation, practices, and aftermath of war), the female body in the Swedish sample can be more easily assigned to one of these dimensions than the female body in the Swiss sample. There are various factors in the Swedish case pleading for the dimension *preparing for war*: women usually appear fully equipped and armed and are shown in active outdoor contexts. The tasks that women are carrying out in the Instagram pictures point to the main task of the Swedish military, which is territorial defence, thus clearly portraying women in a setting that is associated with the organisational identity of the Swedish Armed Forces. In regard to the *preparing* dimension, the female body in the Swedish sample is therefore portrayed as equally strong as that of male soldiers, thus appearing adjusted to the military organisation. There are also examples of women taking part in rather traditional *practices* of war, such as being pictured aboard naval ships and in the context of international operations. The images in the Swedish sample suggest that there is little room allowing women to highlight (stereotypical) femininity, and that such highlighting is not customary. The findings of the Swedish case study could be interpreted as female soldiers being de-feminised, which corresponds to the findings of previous research. In order for women to fit into the organisational identity, they are embodied as 'military men'.

In the Swiss sample, we found fewer associations with the military organisation in the pictures showing women. Even though more than half of the women are portrayed wearing a camouflage suit, we found no scene in the Swiss case portraying women in an active outdoor environment that could be associated with military training or combat skills. There is only one picture of a woman holding a weapon and it is lacking presumed military elements like discipline and order. Moreover, nothing in the surroundings or in the backgrounds hints to the photos being taken at a training base or in another environment which is typical for the armed forces and would mark the difference to other organisations. The images in the sample are therefore neither representing women *preparing for war* nor in the context of the *practices of war* according to McSorley's dimensions, and they do not apply to the third category – *aftermath of war* – either. This suggests that in the Swiss Armed Forces, women are not seen as connected to the core identity of the military, i.e. the defence of the country.

Table 1 Presence of smile, animals, and weapons in Swedish and Swiss sample

	Sweden (n = 45)	Switzerland (n = 28)
Smile	18%	43%
Animals	4%	14%
Weapons (systems)	64%	4%

Table 2 Type of clothing in the Swedish and the Swiss sample

	Sweden (n = 45)	Switzerland[a] (n = 28)
Camouflage	67%	50%
Other military uniform	17%	0%
Non-military clothing	8%	46%
Not recognisable	8%	7%

[a]There are two pictures portraying two women, one of whom is wearing camouflage, while the other is wearing civilian clothing. Thus, both categories apply, which leads to a total percentage of 103%

Various aspects in the portrayal of women on the Swiss Armed Forces' Instagram account highlight their stereotypical femininity. More than half of the women portrayed in the Swiss sample are smiling. This puts an emphasis on the emotional trait that is already associated with femininity and regarded as a specifically female quality. In the Swedish sample, only 18% of the portrayed women are smiling (see Table 1). Furthermore, women in the Swiss military are without exception portrayed in rather calm and peaceful contexts – often the photos were taken indoors and the assigned tasks are not distinctive to the military. The female body is not associated with strength, physical fitness or as a 'war machine' in these pictures. Therefore, the connection to the typical or ideal soldier is not likely to be established. There is also no indication that such qualities are required for the tasks and functions assigned to the portrayed women. In this regard, the way women are portrayed on the Swiss Armed Forces Instagram account matches the view of the Swiss population concerning women's physical ability for military tasks. Women in the Swedish Armed Forces, on the other hand, can be clearly recognised as soldiers preparing and training for war – not only based on their clothing, but also due to the surroundings and their equipment.

These differences in the portrayal of women in the military – whether this form of communication is consciously pursued or not – are likely to be a reflection of the different recruiting strategies of the Swedish and the Swiss Armed Forces as well as the two countries' diverging policies regarding gender equality. The different starting conditions for men and women in the Swiss Armed Forces, which are based on conscription being limited to men, lead to unequal opportunities (Bondolfi 2012). The portrayal of women on Instagram in the Swiss sample suggests that women are integrated, yet not fully included in the military organisation. This means that while women do appear as members of the Swiss Armed Forces, the way they are portrayed by the organisation nevertheless gives the impression that women are engaged in separate domains and represented only in a small part of the wide spectrum of functions the military has to offer. In the Swedish case, on the other hand, and in accordance with the recruiting system, women appear to be fully included in the Swedish Armed Forces, sharing all domains with their male comrades.

7 Conclusion

The aim of this study was to analyse the way women are portrayed on the official Instagram accounts of the Swedish and the Swiss Armed Forces. Our analysis shows that military organisations as gendered organisations can have different ways of managing the challenge that the inclusion of women poses. Previous research focused on how military women are pictured in films as well as how (real) female soldiers are portrayed in the media. Based on our analysis, we can show that the military organisation itself pursues strategies of either assigning women to stereotypical masculine roles *or* of feminising women as a means to manage them in relation to the organisational identity of the military. What may appear as a politically correct intention of over-representing women in social media as compared to actual figures in the organisation is, in fact, a complex way of managing the challenge women pose in relation to a masculine military ideal. In this regard, we noted that Swiss women do not fit into – or seem not to be allowed to take part in – the mainstream embodiment of the military body as identified in the literature (e.g. McSorley 2013a, b). They are thus rendered invisible as military bodies. In the Swedish case, on the other hand, women embody the ideal soldier, thus appearing as assimilated to the military organisation's identity – albeit somewhat subordinated to hegemonic masculinity.

It is likely that the historic backgrounds of the two countries with regard to gender equality and especially the different recruiting strategies of the Swedish and the Swiss Armed Forces affect the position of women in the military and therefore their portrayal on Instagram. Instagram accounts of other armed forces would have to be analysed with reference to their societal and political context in order to establish patterns in this matter. For future research, we suggest taking into account the portrayal of men on armed forces' Instagram accounts. This was not analysed in this study because the focus and interest were on the differences in the portrayal of women between the two countries' armed forces. In order to fully examine the portrayal of gender, it would be important to include the analysis of men as well, which could shed a different light on women's portrayal and put them into another relation. In addition, the sample could be extended to pictures portraying more than two persons. Furthermore, it would be interesting to explore Instagram posts from previous years and to continue exploring this research area by – as was already mentioned – including further countries into the analysis.

This study not only provides insights into how women are portrayed by military organisations on social media, but also demonstrates the influence visual media may have on how an organisation is perceived. During the analysis and while going through the many Instagram posts – only one of various sources – of the Swedish and Swiss Armed Forces, it became apparent how strongly the military depends on imagery to pursue and convey its organisational identity. This is another aspect that deserves closer attention in future research on (social) media and the military.

References

Acker J (1990) Hierarchies, jobs, bodies: a theory of gendered organizations. Gend Soc 4(2):139–158

Addelston J, Stirratt M (1996) The last bastion of masculinity: gender politics at the citadel. Res Men Masculinities 9:54–76

Alvinius A, Holmberg A (2019) Silence-breaking butterfly effect: resistance towards the military within #metoo. Gender, Work & Organization [online]. [Viewed 1 September 2019]. Available from: https://onlinelibrary.wiley.com/doi/epdf/10.1111/gwao.12349

Andrén-Papadopoulos K (2009) US soldiers imagining the Iraq war on YouTube. Pop Commun 7(1):17–277

Autenrieth UP (2014) Die Bilderwelten der Social Network Sites. Bildzentrierte Darstellungsstrategien, Freundschaftskommunikation und Handlungsorientierungen von Jugendlichen auf Facebook und Co. [The picture worlds of the social network sites. Image-centered presentation strategies, friendship communication and action orientations of youth on Facebook and co] Nomos Verlagsgesellschaft, Baden-Baden

Bondolfi S (2012) Wehrpflicht und Geschlecht. Beschränkung der Wehrpflicht auf Männer aus rechtspolitischer Sicht. [Conscription and gender. Limitation of conscription to men from a legal point of view.] Military Power Revue der Schweizer Armee, 1

Burridge J, McSorley K (2013) Too fat to fight? Obesity, bio-politics and the militarization of children's bodies. In: Mc Sorley K (ed) War and the body: militarisation, practice and experience. Routledge, Abingdon, pp 62–77

Moehlecke de BE, Schneider O, Szvircsev Tresch T (2019) #Inclusion - the impact of social media communication on the legitimacy of armed forces. Contemporary Military Challenges, June 2019. Journal of General Staff of the Slovenian Armed Forces, 43–59

Duncanson C (2009) Forces for good? Narratives of military masculinity in peacekeeping operations. Int Fem J Polit 11(1):63–80

Eidgenössisches Büro für die Gleichstellung von Frau und Mann EBG (2010) Anerkennung und Aufwertung der Care-Arbeit – Impulse aus Sicht der Gleichstellung. [Recognition and upgrading of care work – impulses from the perspective of equality.] Eidgenössisches Departement des Innern (EDI), Fachbereich Gleichstellung in der Familie. [online]. [Viewed 8 August 2019]. Available from: https://www.ebg.admin.ch/dam/ebg/de/dokumente/care/anerkennung_und_aufwertungdercare-arbeit.pdf.download.pdf/anerkennung_und_aufwertungdercare-arbeit.pdf

Godfrey R, Lilley S, Brewis J (2012) Biceps, bitches and borgs: reading Jarhead's representation of the construction of the (masculine) military body. Organ Stud 33(4):541–562

Goffman E (1959) The presentation of self in everyday life. Doubleday Anchor, Garden City

Hale H (2012) The role of practice in the development of military masculinities. Gender work Organ 19(6):699–722

Hearn J (2011) Men/masculinities: war/militarism—searching (for) the obvious connections. In: Kronsell A, Svedberg E (eds) Making gender, making war. Routledge, London, pp 49–62

Hellman M (2016) Milblogs and soldier representations of the Afghanistan war: the case of Sweden. Media War Conflict 9(1):43–57

Hellman M, Wagnsson C (2013) New media and the war in Afghanistan: the significance of blogging for the Swedish strategic narrative. New Media Soc 24:1–18

Hellman M, Olsson E-K, Wagnsson C (2016) EU armed forces' use of social media in areas of deployment. Media Commun 4(1):51–62

Hinjosa R (2010) Doing hegemony: military, men, and constructing a hegemonic masculinity. J Mens Stud 18(2):179–194

Holmberg A (2015) A demilitarization process under challenge? The example of Sweden. Def Stud 15(3):235–253

Holmberg A. and Alvinius, A. (2019). How pressure for change challenge military organizational characteristics. Def Stud 19: 130-148. https://doi.org/10.1080/14702436.2019.1575698

Höpfl H (2003) Becoming a (virile) member: women and the military body. Body Soc 9:13–30

Lawson S (2014) The US military's social media civil war: technology as antagonism in discourses of information-age conflict. Camb Rev Int Aff 27(2):226–245

Lobinger K (2012) Visuelle Kommunikationsforschung. [Visual communication research.] Medienbilder als Herausforderung für die Kommunikations- und Medienwissenschaft. Springer Fachmedien, Wiesbaden

Lundqvist I (2018) Bilden av en krigare: Ett viktigt vapen när stereotyper utmanas. [The image of a warrior: an important weapon when stereotypes are challenged.] Bachelor thesis, Department for Communication and Media, Lund University

McSorley K (ed) (2013a) War and the body: militarisation, practice and experience. Routledge, Abingdon

McSorley K (2013b) War and the body. In: McSorley K (ed) War and the body: militarisation, practice and experience. Routledge, Abingdon, pp 1–31

Muhr SL, Slok-Andersen B (2017) Exclusion and inclusion in the Danish military: a historical analysis of the construction and consequences of a gendered organizational narrative. J Organ Chang Manag 30(3):367–379

Mutlu CE (2013) Introduction. In: Salter MB, Mutlu CE (eds) Research methods in critical security studies: an introduction. Routledge, London/New York, pp 139–147

Newlands E (2013) Preparing and resisting the war body: training in the British Army. In: McSorley K (ed) War and the body: militarisation, practice and experience. Routledge, Abingdon, pp 35–50

Paladino L (2015) Drei Frauen, drei Zeitalter, drei Werdegänge. [Three women, three eras, three careers.] Zeitschrift armee.ch, 1 (15)

Pin-Fat V, Stern M (2005) The scripting of Private Jessica Lynch: biopolitics, gender, and the "geminization" of the US military. Alternatives 30:25–53

Rössler P (2010) Inhaltsanalyse. [Content analysis], 2nd edn. UVK, Konstanz

Sasson-Levy O (2011) The military in a globalized environment: perpetuating an 'extremely gendered' organization. In: Jeanes E, Knights D, Martin PY (eds) Handbook of gender, work and organization. Wiley Blackwell, pp 391–410

Seewer G (2003) Vom HD zum Gst Of. Frauen in der Schweizer Armee – 1939 bis in die Gegenwart. [From auxilary service to commissioned officer. Women in the Swiss Armed Forces – from 1939 until today.] Info – Frauen in der Armee, 3 (December)

Strand S, Kehl K (2018) A country to fall in love with/in: gender and sexuality in Swedish Armed Forces marketing campaigns. Int Fem J Polit 21(2):295–314

Swedish Armed Forces (2011) Försvarsmaktens riktlinjer för sociala medier. [Swedish Armed Forces' guidelines for social media.] Document number 17, 100:63719

Swedish Armed Forces (2019) Årsredovisning 2018. [Yearly report 2018.] [online]. [Viewed 1 August 2019]. Available from: https://www.forsvarsmakten.se/siteassets/4-om-myndigheten/dokumentfiler/arsredovisningar/arsredovisning-2018/hkv-2019-02-21-fm2018-10243.7-bilagor-oppna-fm-ar-2018.pdf

Swiss Armed Forces (2019) Armeeauszählung 2019. [Personnel report 2019.] Internal document

Szvircsev Tresch T, Wenger A, Ferst T, Pfister S, Rinaldo A (2015) Sicherheit 2015 – Aussen-, Sicherheits- und Verteidigungspolitische Meinungsbildung im Trend. [Security 2015 – Trends in public opinion on foreign affairs, National Security, and Defence Strategy.] ETH Zürich/ Military Academy at ETH Zurich

Szvircsev Tresch T, Wenger A, De Rosa S, Ferst T, Giovanoli M, Moehlecke de BE, Reiss T, Rinaldo A, Schneider O, Scurrell JV (2019) Sicherheit 2019 – Aussen-, Sicherheits- und Verteidigungspolitische Meinungsbildung im Trend. [Security 2019 – Trends in public opinion on foreign affairs, National Security, and Defence Strategy.] ETH Zurich/Military Academy at ETH Zurich

Traue B (2013) Visuelle Diskursanalyse. Ein programmatischer Vorschlag zur Untersuchung von Sicht- und Sagbarkeiten im Medienwandel. [Visual discourse analysis. A programmatic approach to visibility and expressiveness in media change.]. Zeitschrift für Diskursforschung 2(1):117–136

Welland J (2017) Violence and the contemporary soldiering body. Secur Dialogue 48(6):524–540

Wells K (2014) Marching to be somebody: a governmentality analysis of online cadet recruitment. Child Geogr 12(3):339–353

Wood JT (1994) Gendered media: the influence of media on views of gender. Gend Lives Commun Gend Cult 9:231–244

(Dis-)Empowered Military Masculinities? Recruitment of Veterans by PMSCs Through YouTube

Jutta Joachim and Andrea Schneiker

Abstract State militaries are a central, but not the only site for the construction of military masculinities. In this chapter, we examine how Private Military and Security Companies (PMSCs), which increasingly provide security-related services to armed forces and recruit former military employees, partake in the construction of these masculinities through their social media use. Based on a qualitative content analysis of the YouTube recruitment videos aimed primarily at veterans by two major U.S.-based companies – DynCorp International and CACI – we illustrate how these PMSCs, while affirming traditional notions of military masculinity, challenge its traditional meaning as well by fashioning the rival ideal of the 'corporate soldier'. In addition to upgrading the otherwise marginalised masculinities of veterans by allowing them to be hero warriors, disabled, and civilian employees all at the same time, this ideal also boosts the corporate masculinity of these companies and enables them to define themselves as both legitimate contractors and superior security providers.

Keywords PMSCs · Recruitment · Veterans · YouTube

J. Joachim (✉)
Radboud University, Nijmegen, The Netherlands
e-mail: j.joachim@fm.ru.nl

A. Schneiker
University of Siegen, Siegen, Germany

E. Moehlecke de Baseggio et al. (eds.), *Social Media and the Armed Forces*, Advanced Sciences and Technologies for Security Applications,
https://doi.org/10.1007/978-3-030-47511-6_6

1 Introduction

Social media, developed initially for private consumption, is also increasingly being used by security actors, including armed forces, private security companies, humanitarian NGOs, and terrorist networks. Scholars have therefore started to pay attention to the role Twitter, Instagram or YouTube play for these actors in the promotion and legitimisation of their respective causes.[1] However, the way in which gender is implicated in how these new forms of communication are employed has not been a subject of much research thus far.[2] This is surprising for a variety of reasons. Firstly, as Blower (2016, 89) notes, "cyberspace affords the projection of a new range of identity possibilities". Given that users can insert their voices in a much more unmediated fashion than ever before when posting content – they can choose to remain anonymous or make use of sound, visuality or algorithms – identities, including gender, are likely to be asserted or challenged in an unprecedented manner (van der Nagel 2013).

Secondly, the void in the literature on social media is puzzling since, as existing research shows, gender matters for security actors in their appeal to diverse audiences. Feminist scholars in particular have identified the armed forces as sites where military masculinities are constructed, reproduced, and deployed (Brown 2012).[3] Yet militaries are no longer the only constitutive sites of military masculinities, nor are conventional institutional practices alone formative. Instead, Private Military and Security Companies (PMSCs) are increasingly implicated in the (re-)definition of military masculinities. On the one hand, this starts with the promotion of their services, which range from compound and people protection to reconnaissance and combat, to clients, including armed forces (Joachim and Schneiker 2012, 2018). On the other hand, as we illustrate in this paper, PMSCs (re-)construct military masculinities by appealing to and seeking to hire former military personnel, who "define … the employee base of the private military industry" in significant ways (Singer 2003, 76).

Based on a content analysis of YouTube recruitment videos predominantly aimed at veterans by two major U.S.-based PMSCs – DynCorp International and CACI – we detail how these companies affirm traditional notions of military masculinity, while also challenging it with a corporate rival ideal; an ideal which is coupled to business masculinity and allows veterans to be hero warrior, disabled, and civilian employees all at the same time.

The contribution is structured as follows: firstly, drawing on feminist literature, we define the concept of military masculinity and subsequently develop a set of theoretical assumptions regarding the relationship between social media and gender identity politics. Secondly, we present the empirical results of the qualitative content analysis we conducted of the YouTube recruitment videos by CACI and

---

[1] See Schneiker et al. (2018), Shim and Stengel (2017), Geis and Schlag (2017), Bjerg-Jensen (2014).

[2] For exceptions, see, for example, Jester (2019), Shim and Stengel (2017).

[3] See also: Stiehm (1989), Enloe (1983), Kronsell (2005), Morgan (1994).

DynCorp International. Thirdly, we conclude with a summary of the findings regarding the masculinity constructions deployed and circulated by the two PMSCs as well as a discussion of their implications for the literature on social media, PMSCs, and gender.

2 Military Masculinities

Military masculinity is not just relevant in the context of the armed forces, nor is its meaning fixed. Instead, with the growing relevance of non-state actors related to matters of security and the changes that many armed forces are undergoing as a result of, for example, their professionalisation, the sophistication of weapons' systems or increasing out-of-area missions, the scope conditions of this gendered identity and what it connotes are in flux. In line with these developments, we propose a definition of military masculinity which conceives of it as plural and dynamic in character as well as being a product of social construction. On the basis of this definition and drawing on social media studies, we furthermore develop a set of assumptions as to how military masculinity may be deployed by PMSCs when appealing to ex-militaries on YouTube. While one line of research suggests that these companies can be expected to depict such masculinity in a rather traditional and stereotypical manner, other research leads us to assume that they rely on different, multiple, and perhaps even new variants of military masculinity as social media provides ample opportunities to construct identities.

According to gender scholars, military masculinity encompasses a range of possible positions, identities or performances (Connell 2000) because it is shaped by and intersects with other identity-forming categories, including race, ethnicity or class (Higate and Henry 2016).[4] Moreover, the multiple expressions of this masculinity are reflective of power relationships. According to Heeg Maruska (2010, 238), *hegemonic masculinity* "is one type of identity construct, at the top of a hierarchy that includes subordinate masculinities and femininities". By comparison, *subordinate masculinities* are "oppressed, exploited, and subject to overt control by more dominant forms" (Hinjosa 2010, 181). They are again different from *marginalised masculinities*, which "consist of constructions that are neither dominant nor subordinated, but relegated to being dominated by more powerful forms of masculinity even while they receive a greater share of the patriarchal dividends than subordinated masculinities" (ibid.). In addition to mirroring power relations, masculinity constructs are also "intimately intertwined with the institutions in which individuals are embedded" (ibid.). The armed forces are thus still a central, albeit not the only, site for the construction of military masculinity, which offers "men unique resources

[4] See also: Cornwall and Lindisfarne (1994), Higate and Henry (2016), Petersen (2003).

for the construction of a masculine identity" and as such has significantly shaped its definition throughout time (ibid., 180).

Traditionally, hegemonic military masculinity resembled what Elshtain (1995) refers to as the *just warrior* model and equates with "the brave, physically strong, emotionally tough warrior hero" (Woodward and Winter 2004, 289). However, it is also characterized by self-discipline, self-reliance, and heterosexual desire.[5] Owing to changing circumstances (Woodward and Winter 2004), this "warrior masculinity", while still present today, has undergone certain changes (Duncanson 2009, 66). Especially since the end of the Cold War, it has been subject to a "slight feminization through the construction of a tough and aggressive, yet tender-hearted, masculinity" (Niva 1998, 118; Jester 2019). This finds expression, for example, in what Duncanson (2009, 70) refers to as a "peacekeeping masculinity". The latter is linked to "everyday practices" – for example, in the context of peacekeeping missions – "such as building friendships, drinking coffee and chatting" as well as "to bravery and effective soldiering" (ibid.).

Furthermore, "a move towards 'smarter' armed forces, equipped with technologically sophisticated weapons and intelligence systems" (Woodward and Winter 2004, 295) also contributed to the reconstruction of military masculinity where "the possession of professional skills and expertise" are now seen as characteristic markers (Woodward and Jenkings 2011, 258). Finally, the transition of many armed forces from conscription to professional all-volunteer forces has given rise to what Strand and Berndtsson (2015, 233), based on a study of recruitment information from the Swedish and British armies, refer to as the "enterprising soldier". This soldier is no longer solely motivated by the "mental and physical challenge", the "opportunities to travel", "the excitement", "taking risks", and "wanting to make a difference" or "doing something good" which armed forces have to offer, but is also interested in pursuing a career, growing "professionally and personally", and "assuming responsibility" (ibid., 239–243).

While professionalisation, the integration of women into the armed forces, changes in war-fighting technology, and the increase as well as the growing complexities of international (military) interventions have contributed to the (re-)construction of military masculinity, thus far little attention has been paid to the privatisation of security, which is also a more recent trend concerning the armed forces. Since PMSCs, which are a central dimension of this trend, increasingly perform services for armed forces and hire former military personnel, they can be expected to also co-constitute military masculinity. In order to address this void in the literature, we analyse which masculinities these companies deploy and how they define them.

The YouTube recruitment videos by PMSCs aimed at army veterans offer a particularly good window to capture their conceptions of military masculinities. Firstly, such efforts to attract employees involve, as Brown (2012, 152–154) notes, "overt

[5] See, for example, Higate (2003, 2007), Higate and Hopton (2004), Hockey (2002), Padilla and Riege Laner (2002).

image making and an attempt to sell particular pictures of military service" in addition "to advertising that is steeped in ideas about masculinity and femininity". Secondly, ex-militaries are a particularly interesting group because they unite conflicting masculinities within them (Bulmer and Eichler 2017). While they are celebrated for their past heroism during their time in the military – thus matching the hegemonic ideal of military masculinity – they also frequently suffer from physical or psychological injuries and have trouble reintegrating into civilian life, i.e. exhibit characteristics of subordinate (military) masculinities. Therefore, it is of interest to examine on the basis of which masculinities PMSCs appeal to veterans. Finally, social media such as YouTube provides researchers with excellent laboratories as to how masculinities are deployed, since the unique features of these new forms of communication allow for ample opportunities to construct identities (Blower 2016; Davis 2018).

3 Identity Construction and Social Media

Social media provides users in an unprecedented manner with a myriad of ways of how to present themselves online. Contrary to traditional media outlets, platforms such as Twitter, Instagram or YouTube do not require any intermediaries. Instead, they can be auto-controlled and permit users to interact directly with each other (Stier et al. 2018). Furthermore, the special effects, sound, and video functions of many platforms enhance the possibilities for self-identification (Waters and Jones 2011). According to existing research related to gender and social media, the preciseness with which identities are depicted may, however, vary. Some studies find that these virtual platforms encourage the depiction of identities in a rather traditional and stereotypical manner, thus reinstalling and affirming them. Others, by comparison, find that social media lends itself to the reconstitution and transformation of identities.

With respect to the latter strand of scholarly works, post-structural feminists, for example, emphasise that these new forms of communication hold the potential to escape oppressing and limiting gender identities and to construct more egalitarian and empowering ones. Based on the examination of blogs written by women, Blower (2016, 100) finds that "the medium has encouraged women to reformulate processes of self-realization [and] enables them to explore the self as fluid and plural and from multiple perspectives". In a similar vein, Webb and Temple (2015, 640) in their survey on "social media and gender" conclude that these new forms of communication provide "a space of gender liberation where gender can be performed, conceptualized, and theorized in innovative ways".[6] Davis (2018, 2) also recognises the potential of social media platforms, which, compared to "traditional media …, give the user power" insofar as he or she "can use social media to create and

[6] See also: Bailey and Telford (2007), Hans et al. (2011), Loureiro and Ribeiro (2014).

distribute their own self-representations to the public" (ibid.). Studying the use of Instagram by women over forty, Davis (2018, 2) found that the platform provided opportunities for "... resistance to the cultural norm", which dictates that they "should only self-identify as non-sexual mothers and caretakers". Finally, based on her analysis of posts by girls, Senft (2008) considers it possible for these users to upset the dominant, stereotypical definitions of "girl" and perhaps even "challenge gender-based constraints that hinder social equality" (cited in Webb and Temple 2015, 639; see also Bailey et al. 2013).

Although cognizant of the possibilities of identity (re-)construction, scholars also found evidence to the contrary. In the eyes of Ringrose (2011), social media platforms tend to reinforce stereotypical depictions of gender identities. Men are most likely to be shown as dominant, active, and independent, while in the case of women their attractiveness and dependence is emphasised (Rose et al. 2012, cited in Davis 2018). Emmons and Mocarski (2014) arrive at similar conclusions when studying the depiction of athletes on social media platforms. Compared to their male counterparts, who are generally shown in active performance roles and are "more likely to look away from the camera and be in motion", female athletes most often appear in non-active positions and are "more likely to pose for photos and smile" (125). Similarly, Vandenbosch and Eggermont (2015) observe that social media stimulates self-objectification among female users. Moreover, according to Sills et al. (2016), quite often a celebration of male sexual conquest, slut shaming, and the sexualisation of women can be found on different platforms, which may in turn have offline effects, for example by perpetuating a rape culture. Studies examining the role of race and ethnicity on social media are also suggestive of the reinforcement of traditional identity types. Rather than observing a "new" and "color-blind racism", scholars found evidence of an "old" form of racist discourse to have re-surfaced, which "explicitly imputes racial difference and exclusion" (Cisneros and Nakayama 2015, 108).

When following the debate on social media and gender, analyses of whether these forms of communication encourage the re-constitution of gender constructs or, instead, affirm more traditional ones, several things are noticeable. To start with, there appears to be primarily a focus on women and how they are depicted, while men or masculinity are, for the most part, treated as a baseline for comparison rather than a subject of study per se. Furthermore, scholars have mostly examined the social media use by individuals and how they conform to or challenge existing gender norms. However, although potentially fundamental, corporate actors and their depiction of gender roles via social media in the course of, for example, corporate branding or recruitment, are – apart from few exceptions – yet to be addressed in the literature (e.g., Jester 2019).

Against this backdrop, we examined the YouTube recruitment videos aimed specifically at veterans issued by CACI and DynCorp International, two U.S.-based PMSCs. Both companies are contractors of the U.S. government, offering a broad range of security and military-related services (Isenberg 2009; Military Times 2017). Although these PMSCs also appeal to prospective employees through Twitter (Joachim et al. 2018) or offline and through word-of-mouth (Petersohn 2018), we limited our analysis to the videos of both companies aimed at veterans. Unlike other social media channels, they provide a much richer form of data to examine the (re-) construction of masculine identities. Because they are based on audio and sound, YouTube videos are particularly powerful tools for corporate actors to build an "identity and strengthen their relationship with external stakeholders" (Waters and Jones 2011, 253).[7] According to Waters and Jones (2011, 249–253), who examined the use of videos by non-profit organisations, such platforms "put a human face on the organization" and create "a strong mental impression of the organization in the public's mind" because they permit their users "to tell their story in a powerfully, emotionally connecting way".

CACI's video is titled "Deploying Talent – Creating Careers" (CACI 2013). It is 4.46 min long and consists of sequences from interviews with three CACI employees: Denyse, a recruiting manager; Jared, a field engineer; and Stand, a hiring manager. Denyse is a coloured woman, and Jared and Stand are white men. The video starts with stills combined with overlying text regarding the situation of veterans, followed by alternating interview sequences with the respective spokespersons. The DynCorp video, titled "Proudly Employing Those Who Served", consists of a 3.41-min presentation that combines a sequence of interviews with a veteran named Clint with scenes from both his military service and the company's various activities (DynCorp 2012). The video starts with Clint's personal history, beginning with his decision to join the armed forces after 9/11, followed by a description of his experiences serving in Iraq and the injury he sustained while stationed there.

In order to determine which types of masculine identities both companies construct and how they do this, we conducted a computer-assisted, qualitative content analysis of the CACI and DynCorp videos using MAXQDA. Following the transcription of the two videos, we coded sound and visuals separately, which increased the number of observations. We coded deductively based on the attributes that scholars associate with the different hegemonic military masculinities. In consideration of what are deemed traditional types of masculinity, the codes we used included *patriotism*, *comradeship*, *adventure*, *common experiences*, and *honour*.[8] With regard to what researchers consider a corporate variant, the codes included, among others, *benefits* (*salary*, *pension*, etc.), *upward mobility*, and *positive working environment*.[9] In contrast, the codes for the feminised versions of military masculinity ranged from *compassion* and *caring* to *emotional* (Joachim et al. 2018). This deductively generated list of codes was inductively supplemented throughout the analysis by means of modifying or adding codes, especially to capture subordinate military masculinities. These do not figure very prominently in the literature – with the exception of homosexuals or new recruits, who are referred to as 'ninnies'. Both videos by CACI and DynCorp were coded several times by three trained individuals.

---

[7] See also Devereux (2017), Boateng and Okoe (2015).

[8] See, for example, Gareis et al. (2006), Tomforde (2010), Johansen et al. (2014).

[9] See, for example, Strand and Berndtsson (2015), Levy et al. (2007), Eighmey (2006), Joachim et al. (2018).

4 YouTube Recruitment of Veterans by PMSCs and the Construction of the 'Corporate Soldier'

Western societies rarely "offer an established manner of assessing the veteran experience in union with civil society" after deployment and military service (Gustavsen 2016, 23). PMSCs, however, claim to be able to fill this void and to relate "to what the veterans have been through" (ibid.). Similar to "settled cultures", where the return of soldiers from military conflict or their societal presence is a more frequent occurrence, these companies promise ex-militaries "a distinct repertoire of established cultural resources to frame the[ir] experiences" in the form of masculinities (ibid., 32). They boost the subordinate masculinities of physically and psychologically impaired veterans by affirming their former hegemonic military masculinity and by coupling it with business masculinity, which ex-militaries are promised to acquire upon their entry into the private sector. Therefore, due to the identity (re-) construction involved, online recruitment of former military personnel is not only a functional necessity for PMSCs, but also politically consequential and indicative of broader identity shifts in state-society relations (Kronsell 2005).

4.1 Upgrading Marginalised Masculinities Through the Affirmation of Traditional Military Masculinities

According to Gustavsen (2016, 21), it is important for veterans to "find positive meaning in [their] experiences" in a conflict zone, especially if they return with physical and emotional injuries. Both CACI and DynCorp International provide this positive affirmation with respect to the current status and the past experiences of ex-militaries. CACI is most outspoken in this respect. Referring to itself with feminine attributes such as caring and emphatic and as being cognizant of veterans' marginalised masculinities, the company prides itself on being especially understanding of "the vulnerabilities many service members feel when they realize that the transition from the military to the civilian sector becomes a reality" (CACI 2013, 1.01–1.13). Acknowledging that over "320,000 of our nation's 1.6 million deployed service members have suffered from traumatic brain injury" (CACI 2013, 0.03–0.09) and that the "unemployment rate for wounded warriors is 50 % higher" compared to "the average citizen" (CACI 2013, 0.15–0.19), CACI proudly states that it has "successfully hired hundreds of disabled veterans and wounded warriors into the organization" (CACI 2013, 2.20–2.24).

In much the same manner, DynCorp asserts that it "is committed as a whole to supporting wounded warriors" (DynCorp 2012, 2.59–3.02). At first sight, these statements by the two PMSCs do not appear to be in line with the findings in the social media literature. Rather than being presented in a stereotypical manner as strong and courageous, veterans are depicted with reference to their injuries and frailties. These are, however, not perceived as impediments, but instead as assets from which the companies can profit not only with respect to their public image, but also in terms of human capital.

Although hiring veterans is not unusual and is common in other sectors as well, former members of the armed forces – whether they be "ex-Green Beret, ex-Paratrooper, ex-General, and so on" – are particularly sought-after employees in the private security industry (Singer 2003, 76). It is not only their skill-set and their training that makes them a perfect match for the security industry, as is conventionally claimed by industry representatives (Ramos 2013), but also, as the YouTube videos suggest, their multiple masculinities. In addition to acknowledging that ex-militaries are vulnerable and needy (Ortiz 2012), both PMSCs also affirm their traditional military masculinities based on heroism and bravery. The DynCorp video shows soldiers carrying a stretcher with a wounded comrade and saluting in front of the American flag, while former veteran and now employee Clint shares his traumatic experiences from his deployment, including an incident where "another Marine, sitting next to [him], was killed in the blast" (DynCorp 2012, 1.06–1.14). Hence, compared to other civil employment sectors, where the masculinities veterans acquired in the armed forces do not fit neatly with what is required on the job and therefore requires them to retrain (Ramos 2013), the private security industry does not expect veterans to strip themselves of their military identities. Instead, DynCorp encourages veterans to "continue [with] the same dedication to service and honor and values" (DynCorp 2012, 3.08–3.13). Similarly, CACI honours the "proven leadership, values, job skills, and work ethic" of its prospective employees (CACI 2017) as well as their "brother mentality" (CACI 2013, 3.45–3.52). The fact that ex-militaries can join these companies as they are without having to break with their past is, in addition to such statements, also conveyed through the personal narratives in the videos by the two PMSCs.

According to Denyse from CACI, everything she "learned in uniform" can be "mastered during [a] career with CACI" (CACI 2013, 2.07–2.14). Clint of DynCorp praises his company for upholding the same values he was taught during his military service, including "principles", "honour", "values", "commitment and dedication" (DynCorp 2012, 2.34–2.38). He continues by stating that "we talk about leadership in DynCorp. It's part of our corporate dialogue. And I think it's important. I know from my experiences in the Marine Corps it's all about principle and order and values and dedication" (ibid., 2.26–2.37). Clint's and Denyse's personal observations regarding the fact that their military identities can be linked unproblematically with their identities as civilian employees is, on the one hand, given further meaning through imagery. For instance, in DynCorp's recruitment video, employees are depicted next to and working as a team with U.S. soldiers, helping them lift cargo or repair a machine (DynCorp 2013, 1.50–1.57). On the other hand, the linking of these two identities is reinforced through language, for example by the military jargon with which CACI addresses veterans and affirms their military identity: "In [the] Air Force, we say 'Check 6! Who has your back?' You know, if you've served, you understand what it means to have somebody's back. So, imagine that […] a veteran is now a CACI employee – that person has your back. You know? They do!" (CACI 2013, 3.55–4.14).

Based on their YouTube videos, both PMSCs conceive of military masculinity as an integral part of and compatible with veteran's employment in the security industry. The skills and values that ex-militaries acquired in the armed forces are as valuable and essential for these companies as are the stories and myths surrounding them – that is, the romanticising and the "significant status" enjoyed by those "who have served (honorably)" (Dandeker et al. 2006, 164).

4.2 Coupling and Upgrading Military Masculinity with Business Masculinity

At the same time as they are affirming the military masculinity of veterans, PMSCs are also redefining and further upgrading this masculinity by linking it to civilian and especially business masculinity. Working for a PMSC allows for both: "the 'possibility, application and control of violence' – which used to be also 'the very condition for military service'" – but also to pursue "self-fulfilment, self-enterprise, and personal growth" (Strand and Berndtsson 2015, 234). Rather than merely being a soldier motivated by values, such as a sense of duty, veterans become 'corporate soldiers' who are engaged in an enterprising activity equivalent to that of a business person.

Studying the treatment of veterans in Sweden, Strand (2018, 6, in reference to Duncanson 2009) observes that the construction of soldiers and veterans as "masculine warriors" is "complemented with, and perhaps challenged by," other forms of identity. This also applies to ex-militaries who are recruited by PMSCs. Their "military identities" are "produced by, and reproductive of, a neoliberal regime of government, enacted through rationalities and techniques of the market" (ibid., 3). The statements by Jared of CACI are illustrative in this respect. When "looking for a job" he could do and a company "that would hire military people", CACI was a place that not only offered him "a lot of career path choices", but also valued him as someone who had "served in the U.S. military in Iraq" (CACI 2013, 0.25–0.30), thus acknowledging the "very valuable experience that [many of its employees] had" (ibid., 4.15–4.18).

Clint of DynCorp, a wounded and disabled veteran, equally felt that DynCorp appreciated both "his experiences and [his] limitations", welcoming him "with open arms" (DynCorp 2012, 2.50–2.57). Thus, when he started working for the company, he was certain that he "was in the right place" (DynCorp 2012, 2.38–2.29). Statements such as these can be found throughout the YouTube videos by CACI as well as by DynCorp and suggest that the business masculinity veterans acquire is and remains tightly linked to military masculinity. As was already illustrated in the previous section, the pursuit of a civilian career with a PMSC such as CACI or DynCorp does not require veterans to let go of their past. Instead, the "meaningful job opportunities in fields closely aligned with their military occupational specialties" (CACI 2017) or the "great new careers in technology" as well as "exciting opportunities for experienced professionals, college graduates, and vets" – as CACI also emphasises on other social media channels such as Twitter (Joachim et al. 2018, 305) – build upon the skill-set and the values these ex-militaries acquired in the armed forces.

Although CACI and DynCorp are only two, albeit market-leading, companies in the booming and growing security market, reflections of industry representatives suggest that the coupling of the business and the military identity matters beyond these two illustrative cases. In an article published in the journal *Security*, for example, Jerold Ramos, a U.S. Navy veteran lists several reasons as to why, in his view, veterans are desired employees of the private security industry. In addition to training, "high-tech experience with sophisticated systems and software", and "adaptability" when it comes to "evolving circumstances", these reasons include veterans' ability to "transition quickly from one assignment or focus to another", their "leadership qualities", their "commitment to service", and their "dependability" (Ramos 2013, np). Furthermore, according to White (2017, 14), PMSCs associate "reputational benefits" and "human resource functions" with these former members of the armed forces.

5 Conclusion

In scholarly literature, veterans, given their past, are quite frequently conceived of in a dichotomous fashion as either being in need of protection or as warrior heroes (Kronsell 2012; Dyvik 2016; Åse and Wendt 2018). Our analysis of PMSCs online recruitment campaigns aimed at ex-militaries suggests that such a conception of their masculinities needs to be revised when it comes to the private security sector. Rather than being kept separate, the two identities are reconstructed and become tightly linked when former military employees transition from the armed forces to PMSCs. While showing appreciation for the marginalised masculinities especially of disabled and emotionally traumatised veterans, PMSCs such as CACI and DynCorp promise these individuals a viable and superior alternative. When working for such companies, ex-militaries acquire a business identity and become productive civilians, while at the same time being able to retain the identity of their past as honourable soldiers.

These findings are important in several respects. To begin with, they lend force to the critical security studies literature, according to which PMSCs are neither apolitical nor mere service providers, but are instead actively involved in the ongoing political struggles over the definition of masculinities (Joachim and Schneiker 2012, 2015, 2019). Furthermore, our analysis reveals that social media holds infinitely more possibilities for such companies to (re-)construct identities. Compared to traditional channels of communication, social media allows for identities to be, on the one hand, constituted in more vivid and colourful ways, while, on the other hand, also being linked in new and unfamiliar – or even more ambivalent – manners (Waters and Jones 2011).

In the case of the PMSCs in our study, the coupling of traditional military and business masculinities gives rise to a new 'corporate soldier' identity, which ex-militaries acquire upon their entry into the private security sector. This compound identity, however, is not just consequential for job-seeking veterans. It also is a source of power for PMSCs and an opportunity to reconstruct and upgrade their own corporate identity. By hiring former generals, navy seals or soldiers, PMSCs can, on the one hand, acquire a hegemonic, military masculinity and establish themselves as more legitimate and compatible security actors when seeking contracts with governments and their armed forces. On the other hand, when joining military masculinity with its business variant, these companies can also generate themselves as superior to the armed forces. With the 'corporate soldiers', PMSCs are able to claim to restore veterans to civilians and take care of their economic well-being at the same time as they ensure and provide protection for citizens in general, both from internal and external threats.

Regarding the literature on social media and the construction of identities, the analysis of PMSCs adds to the existing research. Rather than either affirming or transcending gender constructions, these new forms of communication appear to allow for many more possibilities to represent the self and the other. Users may, as is the case for PMSCs, mix and match masculinities or femininities as they see fit, depending on the purpose and the addressees. Moreover, with the rise of social media, the construction of gendered meanings is no longer as tightly linked to official or traditional institutions. Instead, due to the possibilities created by virtual online spaces, private actors are increasingly implicated, thus becoming able to (re-) define what gendered identities connote and, in turn, privilege some and marginalise others.

These insights warrant further scholarly attention and research. As there is still rather little knowledge about how security actors – as opposed to armed forces – (re-)constitute military masculinities or even deploy femininities, future studies may investigate which other variants are brought into circulation by these actors, how they are strategically used, and how states and other actors respond to them. Furthermore, as social media is as yet a rather underexplored empirical domain as far as the identity construction of security actors is concerned, it is important not only to map the kinds of gendered identities that are deployed by these actors, but also to determine whether they promote consistent or rather different and contradictory constructions across different platforms. Finally, much emphasis has thus far been placed on supplying and creating gendered identities. By comparison, there is close to no knowledge as to how addressees receive and respond to these identity conceptions. On the basis of social media data, however, it is in part possible to start accounting for these lacunae. Given the ability of users to comment on as well as state their own opinions, we can examine whether certain gender conceptions find more or less acceptance as well as trace whether they are gaining traction. In conclusion, much remains to be done with respect to researching the construction of political identities since with the arrival of social media a new variable has come into play.

References

Åse C, Wendt M (2018) Gendering the new hero narratives: military death in Denmark and Sweden. Coop Confl 53(1):23–41

Bailey J, Telford A (2007) What's so cyber about it: reflections on cyberfeminism contribution to legal studies. Can J Women Law 19:243–272

Bailey J, Steeves V, Burkell J, Regan P (2013) Negotiating with gender stereotypes on social networking sites: from "bicycle face" to Facebook. J Commun Inq 37(2):91–112

Bjerg-Jensen R (2014) Managing perceptions: strategic communication and the story of success in Libya. In: Engelbrekt K, Mohlin M, Wagnsson C (eds) Lessons from Libya: NATO's 2011 military campaign in context. Routledge, London, pp 171–194

Blower L (2016) It's 'Because I am a Woman': realizing identity to reconstruct identity for the female autobiographical inquiry. Convergence: Int J Res New Med Technol 22(1):88–101

Boateng H, Okoe AF (2015) Consumers' attitude towards social media advertising and their behavioral response: the moderating role of corporate reputation. J Res Interact Mark 9(4):299–312

Brown MT (2012) "A woman in the army is still a woman": representations of women in US military recruiting advertisements for the all-volunteer force. J Women Polit Policy 33(2):151–175

Bulmer S, Eichler M (2017) Unmasking militarized masculinity: veterans and the project of military-to-civilian transition. Crit Mil Stud 3:161–181

CACI (2013) Deploying talents – creating careers program. [Online]. YouTube. [Viewed 29 March 2016]. Available from: https://www.youtube.com/watch?v=UhVXNcADD7c

CACI (2017) Veteran hiring. [Online]. CACI International Inc. [Viewed 17 February 2017]. Available from: http://careers.caci.com/page/show/veterans

Cisneros JD, Nakayama TK (2015) New media, old racisms: twitter, Miss America, and cultural logics of race. J Int Intercult Commun 8(2):108–127

Connell RW (2000) Arms and the man: using the new research on masculinity to understand violence and promote peace in the contemporary world. In: Breines I, Connell RW, Eide I (eds) Male roles, masculinities and violence. UNESCO, Paris, pp 21–33

Cornwall A, Lindisfarne N (1994) Dislocating masculinity: gender, power and anthropology. In: Cornwall A, Lindisfarne N (eds) Dislocating masculinity: comparative ethnographies. Routledge, London/New York, pp 11–47

Dandeker C, Wessely S, Iversen A, Ross J (2006) What's in a name? Defining and caring for "veterans": the United Kingdom in international perspective. Armed Forces Soc 32(2):161–177

Davis SE (2018) Objectification, sexualization, and misrepresentation: social media and the college experience. Soc Med Soc 4(July–September):1–9

Devereux L, Melevar TC, Foroudi P (2017) Corporate identity and social media: existence and extension of the organization. Int Stud Manag Organ 47(2):110–134

Duncanson C (2009) Forces for good? Narratives of military masculinity in peacekeeping operations. Int Fem J Polit 11(1):63–80

DynCorp (2012) Proudly employing those who served. [Online]. YouTube. [Viewed 29 March 2016]. Available from: https://www.youtube.com/watch?v=CvQYDT7y0Zw

Dyvik SL (2016) 'Valhalla rising': gender, embodiment and experience in military memoirs. Secur Dialogue 47(2):133–150

Eighmey J (2006) Why do youth enlist? Identification of underlying themes. Armed Forces Soc 32:307–328

Elshtain JB (1995) Women and War. Basic Books, New York

Emmons B, Mocarski R (2014) She poses, he performs: a visual content analysis of male and female professional athlete Facebook profile photos. Vis Commun Q 21(3):125–137

Enloe C (1983) Does khaki become you? The militarization of women's lives. Pandora, London

Gareis SB, Haltiner K, Klein P (2006) Strukturprinzipien und Organisationsmerkmale von Streitkräften. In: Gareis SB, Klein P (eds) Handbuch Militär und Sozialwissenschaft, 2nd edn. Wiesbaden, VS Verlag, pp 14–25

Geis A, Schlag G (eds) (2017) Visualizing violence: aesthetics and ethics in international politics. Glob Discourse 7(2–3):193–200
Gustavsen E (2016) The construction of meaning among Norwegian Afghanistan veterans. Int Sociol 31:21–36
Hans ML, Lee BD, Tinker KA, Webb LM (2011) Online performances of gender: blogs, gender-bending, and cybersex as relational exemplars. In: Wright KB, Webb LM (eds) Computer mediated communication in personal relationships. Peter Lang Publishers, New York, pp 302–323
Heeg MJ (2010) When are states hypermasculine? In: Sjoberg L (ed) Gender and international security: feminist perspectives. Routledge, New York, pp 235–255
Higate PR (ed) (2003) Military masculinities: identity and the state. Praeger, Westport
Higate PR (2007) Peacekeepers, masculinities, and sexual exploitation. Men Masculinities 10(1):99–119
Higate PR, Hopton J (2004) War, militarism and masculinities. In: Connell RW, Hearn J, Kimmel M (eds) The handbook of studies on men and masculinities. Sage, Thousand Oaks, pp 432–447
Higate P, Henry M (2016) Engendering (In)security in peace support operations. Secur Dialogue 35(4):481–498
Hinjosa R (2010) Doing hegemony: military, men, and constructing a hegemonic masculinity. J Men's Stud 18(2):179–194
Hockey J (2002) "Head down, Bergen on, mind in neutral": the infantry body. J Polit Mil Sociol 30:148–171
Isenberg D (2009) Shadow force: private security contractors in Iraq. Praeger, Westport
Jester N (2019) Army recruitment video advertisements in the US and UK since 2002: challenging ideals of hegemonic military masculinity? Media, War & Conflict [online]. Available from: https://doi.org/10.1177/1750635219859488
Joachim J, Schneiker A (2012) Of "true professionals" and "ethical hero warriors": a gender-discourse analysis of private military and security companies. Secur Dialogue 43:495–512
Joachim J, Schneiker A (2015) The license to exploit: PMSCs, masculinities, and third-country nationals. In: Eichler M (ed) Gender and private security in global politics. Oxford University Press, Oxford, pp 114–130
Joachim J, Schneiker A (2018) Private security and identity politics ethical hero warriors, professional managers and new humanitarians. Routledge Press, London/New York
Joachim J, Schneiker A (2019) Private security and identity politics: ethical hero warriors, professional managers and new humanitarians. Routledge Press, London
Joachim J, Martin M, Lange H, Schneiker A, Dau M (2018) Twittering for talent: private military and security companies between business and military branding. Contemp Secur Policy 39(2):298–316
Johansen RB, Laberg JC, Martinussen M (2014) Military identity as predictor of perceived military competence and skills. Armed Forces Soc 40(3):521–543
Kronsell A (2005) Gendered practices in institutions of hegemonic masculinity: reflections from feminist standpoint theory. Int Fem J Polit 7(2):280–298
Kronsell A (2012) Gender, sex, and the postnational defense: militarism and peacekeeping. Oxford University Press, New York
Levy O, Taylor S, Boyacigiller NA, Beechler S (2007) Global mindset: a review and proposed extensions. Adv Int Manage 19:11–47
Loureiro SMC, Ribeiro L (2014) Virtual atmosphere: the effect of pleasure, arousal, and delight on word-of-mouth. J Promot Manag 20(4):452–469
Military Times (2017) Best for vets: employers 2017. [Online]. Best for Vets. [Viewed 25 October 2017]. Available from: http://bestforvets.militarytimes.com/best-employers-for-veterans/2017
Morgan DHJ (1994) Theater of war: combat, the military, and masculinities. In: Brod H, Kaufman M (eds) Theorizing masculinities. Sage, Thousand Oaks, pp 165–182
Niva S (1998) Tough and tender: new world order masculinity and the Gulf War. In: Zalewski M, Parpart J (eds) The 'man' question in international relations. Boulder, Westview, pp 109–128
Ortiz SR (2012) Veterans' policies, veterans' politics: new perspectives on veterans in the modern United States. University Press of Florida, Gainesville

Padilla PA, Riege Laner M (2002) Trends in military influences on army recruitment themes: 1954–1990. J Polit Mil Sociol 30:113–133
Petersen A (2003) Research on men and masculinities: some implications of recent theory for future work. Men Masculinities 6(1):54–69
Petersohn U (2018) The force of relationships: the influence of personal networks on the market for force. [Online]. European Security. Available from: doi:https://doi.org/10.1080/09662839.2018.1425296
Ramos J (2013) 10 ways military veterans are ideal for physical security sector. [Online]. Security. [Viewed 12 December 2017]. Available from: https://www.securitymagazine.com/articles/84184-ways-military-veterans-are-ideal-for-physical-security-sector
Ringrose J (2011) Are you sexy, flirty, or a slut? exploring 'sexualization' and how teen girls perform/negotiate digital sexual identity on social networking sites. In: Gill R and Scharff C (eds.) New feminities: postfeminism, neoliberalism and subjectivity. London, Palgrave MacMillan, pp 99-116
Rose J, Mackey-Kallis S, Shyles L, Barry K, Biagini D, Hart C, Jack L (2012) Face it: the impact of gender on social media images. Commun Q 60(5):588–607
Senft TM (2008) Camgirls: celebrity and community in the age of social networks. Peter Lang, New York
Shim D, Stengel FA (2017) Social media, gender and the mediatization of war: exploring the German armed forces' visual representation of the Afghanistan operation on Facebook. Glob Discourse 7(2–3):330–347
Sills S, Pickens C, Beach K, Jones L, Calder-Dawe O, Benton-Greig P, Gavey N (2016) Rape culture and social media: young critics and a feminist counter public. Fem Media Stud 16(6):935–951
Singer PW (2003) Corporate warriors: the rise of the privatized military industry. Cornell University Press, Ithaca
Stiehm JH (1989) Arms and the enlisted woman. Temple University Press, Philadelphia
Stier S, Bleier A, Lietz H, Strohmaier M (2018) Election campaigning on social media: politicians, audiences, and the mediation of political communication on Facebook and twitter. Polit Commun 35(1):50–74
Strand S (2018) Inventing the Swedish (war) veteran. [Online]. Critical Military Studies. Available from. https://doi.org/10.1080/23337486.2018.1481267
Strand S, Berndtsson J (2015) Recruiting the "enterprising soldier": military recruitment discourses in Sweden and the United Kingdom. Cri Mil Stud 1(3):233–248
Tomforde M (2010) Neue Militärkultur(en): Wie verändert sich die Bundeswehr durch die Auslandseinsätze? In: Apelt M (ed) Forschungsthema: Militär. Wiesbaden, VS Verlag für Sozialwissenschaften, pp 193–219
Van der Nagel E (2013) Faceless bodies: negotiating technological and cultural codes on Reddit gonewild. J Med Arts Cult 10(2)., np
Vandenbosch L, Eggermont S (2015) Sexualization of adolescent boys: media exposure and boys' internalization of appearance ideals, self-objectification and body surveillance. Men Masculinities 16(3):283–306
Waters RD, Jones PM (2011) Using video to build an organization's identity and brand: a content analysis of nonprofit organizations' YouTube videos. J Nonprofit Publ Sect Market 23:248–268
Webb LM, Temple N (2015) Social media and gender issues. In: Guzzetti B, Lesley M (eds) Handbook of research on the societal impact of digital media. Hershey, IGI, pp 638–669
White A (2017) Beyond Iraq: the socioeconomic trajectories of private military veterans. [online]. Armed Forces & Society. Available from: https://doi.org/10.1177/0095327X17711898
Woodward R, Jenkings KN (2011) Military identities in the situated accounts of British military personnel. Sociology 45(2):252–268
Woodward R, Winter P (2004) Discourses of gender in the contemporary British Army. Armed Forces Soc 30(2):279–301

Part III
Social Media Discussions as Insights into Public Opinion

The Importance of Discussions on Social Media for the Armed Forces

Olivia Schneider

Abstract Social media offers an ideal platform for conducting lively discussions, some of which are also about the armed forces. Therefore, based on theoretical considerations and findings from researching user-generated content, this contribution examines the importance of digital negotiation processes and offers a first theoretical insight into the significance this may have for civil-military relations. A society's image of the armed forces is influenced by the narratives and visualisations conveyed by different media. According to a multi-level model of the public sphere, social media discussions can be understood as part of social negotiation processes. In these processes, a society's image, for example of the armed forces, is consolidated or changed. People participate in discussions on social media for a variety of reasons. On the one hand, cognitive motives are pursued, for instance, to deepen knowledge. On the other hand, it can also simply be a matter of acting out. Negotiation processes on social media should not be neglected, as previous research indicates that online comments influence the attitude and perception of readers. Discussions on social media can thus shape the social vision of armed forces and consequently have an impact on civil-military relations by strengthening or weakening popular support.

Keywords Social media discussions · Social negotiation process · Civil-military relations · Effects of comments on perception

1 Introduction

Social media is characterised, on the one hand, by the fact that the low access barriers make it possible for almost everyone to share content with the digital world, as long as they have internet access and literacy skills. The long time periods spent on

O. Schneider (✉)
Military Academy (MILAC) at ETH Zurich, Birmensdorf, Switzerland
e-mail: oliviaschneider@gmx.net

E. Moehlecke de Baseggio et al. (eds.), *Social Media and the Armed Forces*, Advanced Sciences and Technologies for Security Applications,
https://doi.org/10.1007/978-3-030-47511-6_7

social media[1] as well as the immense amount of content posted on social media illustrate the extent to which these platforms are used nowadays. For instance, per second, 8818 tweets are sent, 966 Instagram photos uploaded, and 1644 Tumblr posts published (Internet Live Stats 2020). On the other hand, social media is also characterised by a social element. "Social action … should mean such action which, in the sense intended by the actor(s), is related to the behaviour of others and is oriented towards this in its course" (Weber 2002, 653). This is also apparent in online follow-up discussions. Readers of comments can use those comments to infer the behaviour of others, which is the basis for action, for example, by reacting in a certain way with one's own comment. In the digital world, readers derive the normative attitudes of others from the attitudes they express and embed in their own comments (Chung 2019).

This contribution focuses on the discussions that take place on social media, investigating the question as to whether such discussions can have an influence on how armed forces are perceived. If simple social interactions on social media can change people's attitudes, it follows that discussions on social media can be of great importance to armed forces, given that these interactions shape the image a society has of them. In order to work out this connection, firstly, the significance of media communication for civil-military relations in a democracy will be explained. Subsequently, the role of social media in this particular context will be discussed.

Social media has made it possible for discussions and conversational structures – which, before the era of social media, were generally conducted in private – to also be conveyed by media and thus become publicly accessible for a large number of people, independent of time and space. This raises the question of why people participate in digital discussions or, more precisely, why comments are made at all. Research has shown that comments are made arising either from cognitive or affective motivation: commentators either wish to broaden their understanding or arguments, or they simply want to let off steam. As of late, the impact of social media comments and discussions has received particular attention, focusing on whether such discussions can actually change views and attitudes on a topic. The findings from this rather young field of research suggest that these effects do, in fact, exist. This indicates that social media discussions should not be neglected, but should be considered an important element in civil-military relations. Therefore, following an overview of the research status, the final step in this contribution is to link and classify online discussions with regard to their significance for armed forces.

[1] In Switzerland, 85% of young people aged 13-19 use social networks daily for entertainment. Around half of the young people use social networks to obtain information (Suter et al. 2018). In the USA, young people aged 8-18 use social media for just over 3 h a day (Common Sense Media 2015).

2 The Role of Media Communication in Civil-Military Relations

Armed forces are integrated into the political system of a society. In established democracies, as part of democratic society they are subject to the same mechanisms as the entire political system and thus must legitimise themselves before the people (Micewski 2013). Legitimacy is gained, on the one hand, by being subject to democratic control. This means that a leadership elected by the people, an independent judiciary, and civil society institutions keep an eye on the armed forces and control them in the interest of the people (Ratchev 2011; Janowitz and Little 1974). In order for this to happen, public access to information is necessary to ensure transparency and respond to the concerns of citizens and the media. On the other hand, armed forces also legitimise themselves by staying current in terms of defence strategies and having appropriate military structures, trained personnel, and modern organisational structures. Under these circumstances, armed forces can carry out the tasks prescribed by policy in an effective, efficient, and socially and financially acceptable manner (Ratchev 2011). Nowadays, modern organisational structures also include the handling and examination of social media. Furthermore, given that legitimacy also means being recognised by citizens as being legitimate and binding (Braun and Schmitt 2009), armed forces, as part of a democratic system, are dependent on the support of the people. One way to maintain contact with the population, gain support in society, and strengthen the reputation of and trust in the organisation is through publicity and communication. An independent political public sphere enables free decision-making and links the state, including the armed forces, with civil society (Habermas 2009). Thus, publicity is an element that makes democracy possible (Martinsen 2008). On the one hand, armed forces are therefore dependent on the media to keep in touch with the population and to inform citizens about the military project, its success, and its costs (Porch 2002). On the other hand, the people, i.e., the sovereign, depend on public reporting on the armed forces in order to keep control over them (Ratchev 2011).

The importance of public communication can be observed, for example, in a study on the perceived effectiveness of the police in Kansas City by Ho and Cho (2017). The study shows that police performance was assessed more positively when there was active communication – even if the crime rate actually increased during that time (ibid.). Accordingly, communication is important for state organisations and security-generating institutions because it enables them to legitimise themselves. Through mediated narratives and visualisations, the mass media shape the social construction of the image of the armed forces present in a society (Virchow 2012; Rukavishnikov and Pugh 2018). Hence, the image that society has of the tasks, legitimacy, structure, and problems of the armed forces is influenced directly by the mass media (Virchow 2012). This applies to a country's own armed forces as well as to armed forces of other states.

Nowadays, social media with its lively discussions is also part of the mass media. It can therefore be assumed that it also influences the image of armed forces in society. Information or opinions on facts and circumstances, for instance regarding the armed forces, are shaped on the basis of direct experiences or through media mediated experiences (Gerhards and Schäfer 2007). Often, only a minority of the population has direct and current experience with or knowledge of an organisation like the armed forces. The majority depend on experiences and knowledge conveyed by the media. However, the media system has changed significantly because of the social and technological changes over the past decade. The development of the interactive Web 2.0 as well as the constant accessibility via mobile internet and, not least of all, social media have reshaped the way in which we inform ourselves and communicate. The effects of these changes will be discussed in the following.

3 Social Media as Simple Public Spheres in the Media Space

While two decades ago, traditional media held the role of gatekeepers, with professional journalists deciding what was being reported on, this has changed with the advent of the internet and especially social media. Nowadays, there are almost no access barriers to publish anything. However, the achieved reach and the generated attention can be very different depending on the medium, account or content (Hendricks and Vestergaard 2019). Armed forces also take an active part in social discourse and proactively communicate with the people through various channels, such as media releases, websites, various social media accounts, and other platforms (Virchow 2012). While social media always poses a certain risk for organisations like the armed forces – for example, too much information could be leaked, which may put soldiers at risk –, it also presents opportunities. This contribution thus focuses on the chances that social media can offer armed forces.

First of all, the internet and social media enable armed forces to publish their own specially prepared content, which can potentially reach a very wide audience very quickly, even without the support of traditional mass media. Another opportunity of social media lies in its interactivity. Social media can potentially not only reach a very large number of people, but one can also enter into a dialogue with them. Therefore, communication is no longer one-sided to a supposedly silent audience, but occurs in direct interaction with the people, who can react publicly and with almost no access barriers. Social media thus enables armed forces to start a dialogue with citizens. This corresponds with new approaches to administrative communication, which state that the people should be supported in being able to articulate shared interests in a pluralistic society. The aim of this type of administrative communication is to promote participation, democracy, good governance, and a better society (Deverell et al. 2015).

As was explained above, the importance of publicity and an interactive, public dialogue with citizens is evident. In the following, the meaning of publicity in this context will therefore be discussed in more detail. To this end, the multi-level model developed by Elisabeth Klaus (2017) shall be applied to determine the importance of social media discussions about the armed forces. A broader understanding of public spheres and political communication, as provided for in the Klaus model, makes it possible to include different sub-publics as well as new types of social media communication (Drüeke 2017). The public, according to Klaus (2017), can be understood as an ongoing process "in which the members of a society come to an understanding on how they want to live. This happens essentially on the basis of their life experiences" (Klaus 2017, 22). Taking armed forces as an example, this means that citizens exchange ideas on what the security-generating organisations and institutions of the state should be like. Personal and media-mediated experiences with armed forces are thematised, generalised, and evaluated. Thus, social ideas about the armed forces can emerge, change or fade.

Such an understanding of political public spheres means that security policy is not only made by the political institutions, but that politics are also "a fundamentally ubiquitous mode of behaviour" (Sutor 1998, 45). It therefore becomes clear that such social negotiation processes can take place in varying degrees of complexity, and that such processes influence how armed forces are perceived in society at every level. In *complex public spheres*, information on the armed forces is prepared in a professional and sophisticated manner and passed on to the population via traditional media, websites, and eGovernment as one-way communication. Consequently, access to complex public spheres is limited because there are only a small number of such forums. The discourse in complex public spheres about security policy and armed forces is therefore determined by already established security policy actors and institutions. These actors often have concrete decision-making power and thus are usually the dominant actors in security policy discussions (Klaus 2017; Drüeke 2017; Katzenbach 2017). Moreover, the more complex the public sphere, the greater is its social power and influence on "public opinion" (Klaus 2017, 26).

Societal negotiation processes can also take place in *medium-sized public spheres*. "Prototypes for medium-sized public spheres represent associations and citizens' initiatives", which can take place off- or online (Klaus and Drüeke 2017, 117). On the internet, however, new forms of communication can also emerge, such as social movements. Examples for this are the Occupy movement, the protests in Hong Kong in recent years or the #FridaysforFuture climate movement, all of which were largely formed via social media. On the one hand, interpersonal communication, such as in follow-up discussions, is important. On the other hand, different roles are developing, and a distinction can be made between active and rather passive members of such movements (Klaus and Drüeke 2017). As a result, certain roles – such as speaker, moderator or audience – emerge, which are, however, interchangeable. They structure the public spheres by steering communication and thus giving it a form (Katzenbach 2017). Hence, opinions can be more easily synthesised, bundled, and made available to a wider public. On the internet, which is understood as a political space, interest groups and social movements can therefore constitute both of state and civil-societal actors (Drüeke 2017).

At the beginning of 2020, for example, a communication space opened up in social media under the hashtag #ww3 and went viral. In a very short time, this hashtag reached several 100,000 people on Instagram, Twitter, and Tiktok, respectively. Using this hashtag, young people in particular published their concerns, views, and statements on a potential third world war, following the tensions between the USA and Iran. Moreover, attitudes towards the armed forces, one's own conscription or a possible deployment in Iran were discussed. This was done in an ironic and cynical way – often in the form of memes or short videos – which, in turn, was taken up, thematised, and criticised by the mass media (Kedves 2020; Alfonso 2020). It is conceivable that a movement could organise itself under this hashtag if concerns in this regard continue. This could happen, for instance, if the individually experienced fear and the perceived need to do something evolves into a collective perception of the problem through diverse and mutual references (Katzenbach 2017). In other words, statements and views on this topic can be bundled and solidified until they could eventually spill over into the offline world in the form of strikes or protests.

Finally, there are other social negotiation processes, however, these are often rather simply interaction systems that arise, for example, when people fleetingly meet in offices or bars (Katzenbach 2017, 154). In such *simple public spheres*, there are no structural distinctions between speaker and audience roles. Simple public spheres can also form online, for instance, when users exchange their experiences and opinions about the armed forces in follow-up discussions (Katzenbach 2017). Patterns of interpretation solidify in simple interactions and everyday communication (Schrape 2011). In fact, all communication contributes to the construction of reality by deciding what is taken up and what is left to be forgotten (Luhmann 2017). Special interest is thus being paid to social media and the new interactivity of Web 2.0 and the simple public spheres occurring there. On the internet and especially on social media, discussion and communication spaces are emerging in which citizens without proven expert knowledge can exchange views on security policy and armed forces. On the basis of their own experiences and knowledge, they discursively negotiate social ideas (Drüeke 2017) and can thus shape, reinforce or change the way the armed forces – including their narratives and visualisations – are perceived in society. In contrast to traditional simple public spheres, however, simple public spheres online are mediated by the media and thus exist independently of time and space. Therefore, as the communication space is generated by media-mediated and publicly accessible contributions, the follow-up discussion loses its fleeting character and takes on a manifest, permanent form (Katzenbach 2017).

These different levels of publicity are not detached from one another, but are mutually interlinked. Mass media are complexity-reducing distribution points that observe sub-publics and bring generally relevant information and views into the public discourse (Marcinkowski 2002; Schrape 2011). Complex public spheres thus have the opportunity to guide and influence social negotiation processes at the less complex levels. In turn, new patterns of interpretation and innovative ideas, for example about armed forces, can emerge in medium-sized public spheres in which interest groups join forces, for instance via social media. In medium-sized public spheres, topics and patterns of interpretation from simple public spheres are bundled and can thus be taken up by the mass media (Katzenbach 2017; Drüeke 2017; Klaus 2017; Schrape 2011). In simple public spheres on social media, alternative patterns of interpretation can also be publicly discussed. Over time, a debate can emerge in which the individual experiences and views of these public spheres are perceived as group experiences and group opinions, and as such can flow into a mass media public sphere (Katzenbach 2017).

Mass media and social media communication serve to construct reality on various levels. At the level of simple and middle public spheres, the patterns of interpretation are more diverse than at the level of complex public spheres. In complex public spheres, comprehensive descriptions of the present and patterns of interpretation are taken up in order to offer society a complexity-reducing orientation basis for follow-up communications (Schrape 2011). Interpretation patterns from the different levels feed the negotiation processes at the other levels and thus reciprocally influence how issues are discussed at the different levels.

As a result, not only direct communication and mass media communication on part of the armed forces will receive special attention, but also particular simple interaction systems and digital follow-up discussions. In the following, I will discuss why people participate and comment on online simple public spheres at all, and which effect online discussions can have.

4 Motivation to Participate in Simple Online Public Spheres

Online follow-up discussions refer to the total amount of user-generated content, i.e., all comments following media content or a post on social media (Ziegele 2016). Comments, which Ziegele (2016) describes as "post-communicative and asynchronous online connection communication", are published online in writing following digital contributions (Ziegele 2016, 38). This enables a large number of actors to participate in a discussion, for example on the armed forces, and to contribute their experiences and ideas to the discussion, independent of space and time. However, why does anyone even take the time to participate and write comments on social media or other interactive platforms? An analysis by Staender et al. (2019) shows that contributions are more often liked or shared than commented on. The different participation intentions require different degrees of user activation. Liking something only requires a small activation – with a single click, the action is already complete. Sharing, on the other hand, already requires a stronger activation than liking (Staender et al. 2019; Berger and Milkman 2012). By sharing, users actively take over content and distribute it under their own name in their own network. By publishing a comment, finally, one gives others an insight into one's own opinion and therefore reveals and presents oneself to others (Staender et al. 2019).

Emotional and surprising content has a strong influence on the willingness to comment (Staender et al. 2019; Ziegele 2016). Emotionality means that positive and negative content occur, thus differing from neutral and factual content. Surprising content that contains something new and is easy to understand stimulates the willingness to comment, because it is read with greater interest than content that is already familiar and simply activates users' prior knowledge. This, in turn, motivates users to contribute their own perspective (Ziegele 2016). A similar situation can be observed with socially relevant and political-ideological topics that are morally controversial and present different points of view that are being discussed controversially. Such topics stimulate the willingness to comment, because the incompleteness of the topic quasi challenges commentators to "influence the process of social opinion formation … by publicly expressing their own views" (Ziegele 2016, 326). However, the fear of negative reactions to a statement on a controversial topic can rather inhibit the willingness of people with less self-confidence to comment (Ziegele 2016). Further research results indicate that commenting fulfils a certain purpose for those who comment. Moreover, a distinction can be made between the different functions commenting has for users in their directly experienceable life world: commenting is based either on cognitive or on affective motivation (Ziegele 2016).

4.1 Cognitive Functions for Commenting

Both the reading of comments and the commenting itself fulfil cognitive functions. Users want to share their views, information, knowledge and personal opinions on various topics with the public (Ziegele 2016). They see commenting as an opportunity to broaden their knowledge, for instance of armed forces and security policy, in a discussion, to better understand facts, and to learn about other opinions. However, commenting is also a way to refute or teach others. Similar to immediate follow-up discussions in traditional conversations, online follow-up discussions allow media content and the climate of opinion to be cognitively processed and classified, leading to a deeper individual understanding and knowledge – offline as well as online. For those who read online follow-up discussions, the search for information is an important aspect (Springer et al. 2015). In this way, further information, patterns of interpretation, and views can be obtained and compared with one's own point of view. The personal bouquet of arguments can thus be newly compiled, decorated, and expanded for a further follow-up discussion. What is more, comments can build a bridge between the personal and social relevance of a news item (Ziegele 2016). In addition, an insight into society and public opinion is possible. Analyses have shown that in online discussions where the majority of comments deviate from the position of the commented article, readers assume that public opinion is more in line with the position of the commentators than that of the article (Lee and Jang 2010).

4.2 *Affective Functions for Commenting*

However, people do not only have cognitive motivations when commenting. Often, media content is processed ironically through commentary, creating an enjoyable atmosphere similar to offline follow-up discussions. Furthermore, media content can also be commented very emotionally. According to Ziegele (2016), both the act of commenting and the emotional discussion have a cathartic, purifying, and distracting effect. Emotions can thus be reduced, and the commentators can distract themselves from everyday stress. Commenting is therefore no longer necessarily about the actual topic of a contribution or a post, such as the armed forces.

Following the publication of tragic media content, consternation and condolences are often expressed (Ziegele 2016). Unfortunately, in the armed forces, tragic incidents occur time and again, which are then communicated to the people by the media. In October 2019, for example, there was a tragic accident during military training in Georgia. A report by ABC News, which was uploaded to YouTube, was commented on more than 1000 times, the majority of those who commented expressing their deepest condolences and dismay to the relatives via YouTube (ABC News 2019).

In this regard, readers of online follow-up discussions are often confronted with quite emotional comments, which are not always about the matter at hand, as commenting is sometimes also used as a valve. However, readers are also confronted with comments that are thoroughly factual. Through such comments, new patterns of interpretation and points of view are conveyed to the readers, while existing ones are confirmed. The effects of this diversity of comments on readers is explained below.

5 Effects of Social Media Follow-Up Discussions

A glance at news sites or social media sites reporting on armed forces topics reveals that information about the armed forces is not only being provided and posted, but that it is also being discussed. This is empirically demonstrated by other contributions in this anthology, such as Kümmel, Leightley et al. or Rivnai. The question arises, however, whether such follow-up discussions have any effect at all on the people who read the comments, or whether online discussions are completely negligible. According to the theory explained above, in such interactions, ideas and patterns of interpretation on various topics are consolidated (Schrape 2011) and should therefore also have an effect on the attitudes of people. In recent years, research has been concerned with and tried to explore the effects of user-generated comments on readers in experimentally designed studies. In order to demonstrate the importance of online follow-up discussions, an overview of the findings of this research is presented below.

Chen et al. (2019) were able to show that negative or positive user comments can influence the purchase intention of consumers. The order of the positive and negative comments plays a role, the latest information dominating the influence on consumers. However, this is moderated by the extent of product involvement, which, as Ho-Dac et al. (2013) noted, is particularly true for products of weak brands. With products of strong brands, online user comments do not have a significant impact on sales. This could mean that follow-up discussions on armed forces could be particularly important if the armed forces have little support and reputation in a society and the brand 'armed forces' is rather weak. Although there is, to my knowledge, no research design on the effect of comments on attitudes towards armed forces as yet, there are findings on other state institutions. An often quoted analysis by Walther et al. (2010) addresses online follow-up discussions on public service announcements (PSAs). PSAs – which are aimed to provide information on health issues and drug prevention and to propagate certain behaviour – are usually initiated by governments and health authorities and are an important part of many government health campaigns. Walther et al. (2010) conclude that the same advertisement is rated better when followed by positive comments than when followed by negative comments. However, the perception of the actual health risk is not significantly influenced by the comments. Nevertheless, a stronger identification with the commenting community leads to an increased effect with regard to the evaluation of the advertisement, as well as with regard to the evaluation of the actual health risk.

The importance of social identification with other commentators was also noted by Chung (2019), whose research indicates that the comments of other users can influence the norms and views on a topic, especially if there is a strong social identification with the commenting community. It is therefore to be expected that the perception of state social media campaigns is also influenced by follow-up discussions. Social media campaigns in the context of armed forces, for example, are often used for the recruitment of young people. However, young people are not only influenced by the campaign itself, but also by the follow-up discussions of their peers on social media. Indeed, the more they identify with the commenting community, the more likely they are to be influenced by the follow-up discussions. For a positive effect of such recruitment campaigns, it is thus likely that it is not only the social media campaign itself that counts, but also the digital follow-up discussion it prompts.

In online discussions, not only is it important whether the comments are positive or negative, but the tone in which comments are written also has a decisive influence. An experimental analysis of the effect of reading online follow-up discussions has shown that the perception and evaluation of a topic is, in fact, not just influenced by the contribution itself (Anderson et al. 2014). Being exposed to rude comments on a topic on the internet (on news sites or social media), for instance, actually influences the perception of the actual topic (Anderson et al. 2014; Jennings and Russell 2019; Lee 2012). The ensuing discussion, in which both polite and impolite comments are made, further shapes attitudes towards a topic, such as security policy or armed forces. In that regard, an analysis of television discussions by Mutz and Reeves (2005) showed that rude political discussions on television reduce trust in institutions. Although, according to Molina and Jennings (2018), uncivil online follow-up discussions receive greater attention, the more polite social media discussions are also more elaborate and lead to greater satisfaction. This was also observed by Jennings and Russell (2019, 428), who found that "civil discussion indirectly predicted an increase in both policy support and intended behavior through attitude formation".

If polite social media discussions lead to greater satisfaction among readers and are more likely to generate political support, such discussions are of interest to armed forces. Should armed forces indeed choose to participate in online follow-up discussions on various interactive platforms, it is important to focus on argumentative and civil discussions. In this way, a positive image can be transported. Support for polite discussions without specific moderation can, for instance, occur through the participating commentators, as politeness in online discussions sets standards of politeness that others are likely follow (Han and Brazeal 2015).

The fact that changes in behaviour and attitudes on political and social issues are not only a short-term effect was demonstrated by Stylianou and Sofokleous (2019). They investigated the influence of online user comments on attitudes towards minorities, concluding that positive comments reduce prejudice towards refugees, a change which still persisted 1 week after the experiment. Such an effect is most pronounced when strong prejudices previously prevailed. "Overall, existing research suggests that online comments on news items affect readers' attitudes, perceptions and evaluations, at least as far as experimental studies can support" (Stylianou and Sofokleous 2019, 129). It is therefore reasonable to assume that comments and follow-up discussions on the internet and social media can also influence the attitudes, perceptions, and evaluations of armed forces, especially if the comments are new, up-to-date, positive, and polite. Moreover, it might be possible to reach people with prejudices against armed forces and perhaps even bring about a more long-term change in their attitude towards armed forces if they read favourable and decent comments on the subject.

6 Conclusion

Social media offers an opportunity for civil-military relations. It enables armed forces to partially observe simple public spheres, which, in turn, makes it possible to follow the social negotiation processes concerning armed forces that take place there. Since such processes used to take place mainly in the private sphere under exclusion of the public, this is a great opportunity made possible by social media. In this way, armed forces can monitor which ideas about them or about security-related issues are becoming entrenched in society. Moreover, thanks to social media, armed forces no longer have to just watch, but now – with all their members – have the opportunity to participate in such social negotiation processes at various levels. As had already been the case before, they can use the complex public sphere and provide professionally prepared, generally relevant information and statements via traditional mass media, such as daily newspapers, television, and radio or government websites. However, they can now also participate through social media on medium-sized and simple public spheres by using the interactivity of Web 2.0 and social media. This applies not only to armed forces as an organisation, but also to military personnel on an individual level. These new ways of participation are an extended possibility of how security-relevant views and interpretation patterns can reach medium and simple public spheres. Thus, armed forces can actively shape their image in society, given the fact that – as commentary research shows – social media discussions influence the attitudes and behaviour of readers.

The tone in which communication takes place is particularly important. Polite online discussions are more argumentatively elaborated and lead to greater satisfaction among readers. Furthermore, the perception of a topic can be influenced particularly strongly if it is a weak brand with little reputation in society. Participation in simple interaction systems is therefore especially interesting for armed forces with a weaker reputation and less support in society. In this way, new and innovative patterns of interpretation of armed forces as well as security-related issues can be discussed. This is especially true when such interactions involve not only laypersons (often only a minority has direct experience with armed forces), but also members of the armed forces with direct experience and expert knowledge, thus complementing the social negotiation processes.

The narratives and visualisations conveyed on different levels in the form of memes or videos also shape the ideas a society has about the armed forces. Armed forces are therefore well advised not only to communicate directly and one-way, but also to interact with the population. It is another chance to present themselves to the people, to legitimise themselves, and to strengthen the support of the citizens. Thanks to the new opportunities created by social media, armed forces as state actors can now also participate in simple public spheres, since these are mediated on social media. Due to the media-mediated element and the written form, this can be done independent of time and space. Armed forces can thus help shape the discourse through tonality and facts, as well as model the image that society has of them beyond the traditional way. This gives social media discussions a special significance for civil-military relations.

This contribution offers a first theoretical insight into social negotiation processes taking place in discussions on social media and into the possible effects that these discussions may have on civil-military relations. There is still much that needs to be done. The considerations made here are based on Elisabeth Klaus' (2017) multi-level model of the public sphere as well as on newer findings from experimental research designs on the effects of user-generated content. In order to examine whether security-related topics and, especially, topics regarding armed forces show similar effects when reading online discussions, future analyses should focus on the effect of positive and negative comments on the perception of and attitude towards armed forces. Further research should also examine empirically how and, more specifically, which topics are being discussed in online discussions about armed forces. Are there differences depending on the account, news site or forum? Are there international differences in the way online discussions about armed forces are conducted? Can connections be established with the reputation of armed forces and the way in which they are being discussed? As this is a very young field of research, there are many questions yet to be examined.

References

ABC News (2019) Military training accident leaves 3 soldiers dead [online]. [Viewed 22 January 2020] Available from: https://www.youtube.com/watch?v=Xl1225Wyy4E [latest update: 22 January 2020]

Alfonso F (2020) How Americans and Iranians are using memes and hashtags to cope with conflict [online]. CNN 2020. [Viewed 12 January 2020]. Available from: https://edition.cnn.com/2020/01/11/middleeast/iran-us-memes-ww3-twitter-trnd/index.html

Anderson AA, Brossard D, Scheufele DA, Xenos MA, Ladwig P (2014) The "nasty effect:" online incivility and risk perceptions of emerging technologies. J Comput-Mediat Commun 19(3):373–387

Berger J, Milkman KL (2012) What makes online content viral? J Mark Res 49(2):192–205

Braun D, Schmitt H (2009) Politische Legitimität. In: Kaina V, Römmele A (eds) Politische Soziologie. Ein Studienbuch. VS Verlag für Sozialwissenschaften, Wiesbaden

Chen H, Yan Q, Xie M, Zhang D, Chen Y (2019) The sequence effect of supplementary online comments in book sales. IEEE Access 7:155650–155658

Chung JE (2019) Peer influence of online comments in newspapers: applying social norms and the social identification model of deindividuation effects (SIDE). Soc Sci Comput Rev 37(4):551–567

Common Sense Media (2015) The common sense census: media use by tweens and teens [online]. [Viewed 25 June 2019]. Available from: https://www.commonsensemedia.org/research/the-common-sense-census-media-use-by-tweens-and-teens

Deverell E, Olsson E, Wagnsson C, Hellman M, Johnsson M (2015) Understanding public agency communication: the case of the Swedish armed forces. J Public Aff 15(4):387–396

Drüeke R (2017) Politische Kommunikationsräume im Internet. In: Klaus E, Drüeke R (eds) Öffentlichkeiten und gesellschaftliche Aushandlungsprozesse. Theoretische Perspektiven und empirische Befunde, Critical Studies in Media and Communication, vol. 14. transcript, Bielefeld, pp 39–62

Gerhards J, Schäfer M (2007) Demokratische Internet-Öffentlichkeit? Ein Vergleich der öffentlichen Kommunikation im Internet und in den Printmedien am Beispiel der Humangenomforschung. Publistik 52(2):210–228

Habermas J (2009) Philosophische Texte. Suhrkamp, Frankfurt am Main

Han S, Brazeal LM (2015) Playing nice: modeling civility in online political discussions. Commun Res Rep 32(1):20–28

Hendricks VF, Vestergaard M (2019) Reality lost. Markets of attention, misinformation and manipulation. Springer Open, Cham

Ho AT, Cho W (2017) Government communication effectiveness and satisfaction with police performance: a large-scale survey study. Public Adm Rev 77(2):228–239

Ho-Dac NN, Carson SJ, Moore WL (2013) The effects of positive and negative online customer reviews: do brand strength and category maturity matter? J Mark 77(6):37–53

Internet Live Stats (2020) [Online]. [Viewed 22 January 2020]. Available from: https://www.internetlivestats.com/one-second/

Janowitz M, Little RW (1974) Sociology and the military establishment. 3rd ed./with [a new] introduction, Sage series on armed forces and society, vol. 6. Sage, Beverly Hills/London

Jennings FJ, Russell FM (2019) Civility, credibility, and health information: the impact of uncivil comments and source credibility on attitudes about vaccines. Public Underst Sci 28(4):417–432

Katzenbach C (2017) Von kleinen Gesprächen zu großen Öffentlichkeiten? Zur Dynamik und Theorie von Öffentlichkeiten in sozialen Medien. In: Klaus E, Drüeke R (eds) Öffentlichkeiten und gesellschaftliche Aushandlungsprozesse. Theoretische Perspektiven und empirische Befunde, Critical Studies in Media and Communication, vol. 14. transcript, Bielefeld, pp 151–174

Kedves A (2020) #ww3 – Junge twittern ihre Angst vor dem dritten Weltkrieg [online]. Tagesanzeiger 2020, 08.01.2020. [Viewed 22 January 2020]. Available from: https://www.tagesanzeiger.ch/kultur/ww3-junge-twittern-ihre-angst-vor-dem-dritten-weltkrieg/story/10343056

Klaus E (2017) Öffentlichkeit als gesellschaftlicher Selbstverständigungsprozess und das Drei-Ebenen-Modell von Öffentlichkeit. Rückblick und Ausblick. In: Klaus E, Drüeke R (eds) Öffentlichkeiten und gesellschaftliche Aushandlungsprozesse. Theoretische Perspektiven und empirische Befunde, Critical Studies in Media and Communication, vol. 14. transcipt, Bielefeld, pp 17–38

Klaus E, Drüeke R (2017) Internetöffentlichkeiten und Gender Studies: Von den Rändern in das Zentrum? In: Klaus E, Drüeke R (eds) Öffentlichkeiten und gesellschaftliche Aushandlungsprozesse. Theoretische Perspektiven und empirische Befunde. Critical Studies in Media and Communication, vol. 14. transcript, Bielefeld, pp 101–126

Lee E (2012) That's not the way it is: how user-generated comments on the news affect perceived media bias. J Comput-Mediat Commun 18(1):32–45

Lee E, Jang YJ (2010) What do others' reactions to news on internet portal sites tell us? Effects of presentation format and readers' need for cognition on reality perception. Commun Res 37(6):825–846

Luhmann N (2017) Die Realität der Massenmedien. Springer Fachmedien Wiesbaden, Wiesbaden

Marcinkowski F (2002) Massenmedien und die Integration der Gesellschaft aus Sicht der autopoietischen Systemtheorie. In: Imhof K, Jarren O, Blum R (eds) Integration und Medien, Mediensymposium Luzern, 7. VS Verlag für Sozialwissenschaften, Wiesbaden, pp 110–121

Martinsen R (2008) Öffentlichkeit in der "Mediendemokratie" aus der Perspektive konkurrierender Demokratietheorien. In: Marcinkowski F, Pfetsch B (eds) Politik in der Mediendemokratie, Politische Vierteljahresschrift, 42, 1st edn. VS Verlag für Sozialwissenschaften, Wiesbaden, pp 37–69

Micewski ER (2013) Werte und Militär – Werte im Militär. Wertethematik im Kontext von Individuum, Gesellschaft und Streitkräften. Österreichische Militärische Zeitschrift (ÖMZ) (3)

Molina RG, Jennings FJ (2018) The role of civility and metacommunication in Facebook discussions. Commun Stud 69(1):42–66

Mutz D, Reeves B (2005) The new videomalaise: effects of televised incivility on political trust. Am Polit Sci Rev 99(1):1–15

Porch D (2002) "No bad stories". The American media-military relationship. Naval War Coll Rev 55(1):85–107

Ratchev V (2011) Civilianisation of the defence ministry. A functional approach to a modern defence institution, DCAF regional programmes series, no 7. DCAF, Geneva

Rukavishnikov VO, Pugh M (2018) Civil-military relations. In: Caforio G, Nuciari M (eds) Handbook of the sociology of the military, Handbooks of Sociology and Social Research, vol 19. Springer, Cham, pp 123–143

Schrape J (2011) Social Media, Massenmedien und gesellschaftliche Wirklichkeitskonstruktion. Berl J Soziol 21(3):407–429

Springer N, Engelmann I, Pfaffinger C (2015) User comments: motives and inhibitors to write and read. Inf Commun Soc 18(7):798–815

Staender A, Ernst N, Steppat D (2019) Was steigert die Facebook-Resonanz? Eine Analyse der Likes, Shares und Comments im Schweizer Wahlkampf 2015. SCM 8(2):236–271

Stylianou S, Sofokleous R (2019) An online experiment on the influence of online user comments on attitudes toward a minority group. Commun Soc 32(4):125–142

Suter L, Waller G, Bernath J, Külling C, Willemse I, Süss D (2018) JAMES Jugend Aktivität Medien – Erhebung Schweiz, Ergebnisbericht zur JAMES-Studie 2018. Hochschule für Angewandte Wissenschaft, Zürich

Sutor B (1998) Politik. Ein Studienbuch zur politischen Bildung, 5th edn. Dr. Paderborn, Schöningh

Virchow F (2012) Militär und Medien. In: Leonhard, N. und Werkner, I.-J. (eds.). Militärsoziologie – Eine Einführung. 2nd updated and completed ed. VS Verlag für Sozialwissenschaften/Springer Fachmedien Wiesbaden GmbH Wiesbaden, Wiesbaden, p 200–219

Walther JB, DeAndrea D, Kim J, Anthony JC (2010) The influence of online comments on perceptions of antimarijuana public service announcements on YouTube. Hum Commun Res 36(4):469–492

Weber M (2002) Schriften. 1894–1922. In: Käsler D (ed) (Kröners Taschenausgabe, 233). Kröner, Stuttgart

Ziegele M (2016) Nutzerkommentare als Anschlusskommunikation. Theorie und qualitative Analyse des Diskussionswerts von Online-Nachrichten. Springer, Wiesbaden

Debating German Special Forces: A Scandal in the Military, a Documentary, and a Thread

Gerhard Kümmel

Abstract In April 2017, a farewell party was held for a company commander of the German special forces (*Kommando Spezialkräfte*, KSK). This party became the focus of a feature by the political television magazine *Panorama* because of the inappropriate behaviour of KSK soldiers, such as a peculiar training course, the singing of extreme right-wing songs, giving the Hitler salute, and – planned, but not performed – sexual intercourse. The feature received numerous comments on *Panorama*'s *Facebook* page. These posts were subjected to an inductive content analysis which identified five different thematic fields: (1) questioning the story, (2) respect and leniency, (3) shock and concern, (4) the Minister of Defence and her discharge of office, and (5) tradition and identity. The posts show a controversial debate, allow for some insight into German civil-military relations, and reveal some implications of the social media age for the armed forces.

Keywords Bundeswehr · Special forces · KSK · Soldierly misbehaviour · Civil-military relations

1 Introduction

The purpose of the present book is to analyse the relationship between the armed forces and social media, and to explore the implications of social media for the armed forces. In order to enrich our knowledge in this field, the following contribution examines the case of a scandal which took place in the German Bundeswehr. It was reported by a German political television magazine and was commented on extensively on the Facebook page of said magazine. The posts in the thread were subjected to a simple inductive content analysis aimed at identifying and structuring

G. Kümmel (✉)
Bundeswehr Center of Military History and Social Sciences, Potsdam, Germany
e-mail: gerhardkuemmel@bundeswehr.org

E. Moehlecke de Baseggio et al. (eds.), *Social Media and the Armed Forces*, Advanced Sciences and Technologies for Security Applications,
https://doi.org/10.1007/978-3-030-47511-6_8

the reactions of the viewers by way of generating different thematic fields based on the empirical material itself. The reactions in the posts were thus grouped into five different categories. This contribution starts by outlining the story of the scandal, before presenting the findings of the inductive content analysis. The findings are then summarised and some conclusions are drawn with regard to civil-military relations in Germany. Finally, some implications of the social media age for the armed forces are illustrated.

2 The Story: The Scandal and the Documentary

On 17 August 2017, the political television magazine *Panorama* on the North German Broadcast (NDR) announced both in a press release and on its Facebook page a documentary on the allegedly inappropriate behaviour of soldiers of the Bundeswehr Special Forces (*Kommando Spezialkräfte*, hereinafter referred to as KSK) and conveyed initial information on the scandal.[1] Later that evening, the nine-minute feature titled "Hitler salute? Investigations against a Company Commander" was broadcast on ARD (Grabler et al. 2017).

The feature tells the following story: In April 2017, the editorial team of *Panorama* was approached by a woman who is referred to as 'Anna' in the documentary. She reports her experiences at a bizarre farewell party in one of the four companies of the KSK. Prior to this party, she is contacted by a KSK soldier, who already knew her from a dating website and seemed to be sure, "that she likes brutal sex with men" (Grabler et al. 2017, 01:15).[2] He tries to persuade 'Anna' to serve as a farewell present and sex playmate for the leaving company commander, Lieutenant Colonel Pascal D., after he mastered a special training course (Gebauer and Lehberger 2017). Financial remuneration is not offered. 'Anna's' participation in the party shall be ensured by promising her the fulfilment of her sexual fantasies. The KSK soldier advertises the characteristics of his boss as follows: "Two meters tall, fucks everything that comes his way. Hands like a toilet lid, and tattooed. In terms of your secret sexual desires this could be the night of your life" (Grabler et al. 2017, 01:22–01:33). He also forwards 'Anna' a video clip which shows his commander practicing kickboxing.

'Anna' accepts the invitation and takes a plane from Hamburg to Stuttgart on 27 April 2017. Two KSK soldiers pick her up and accompany her to a military shooting range close to the city of Sindelfingen (Gebauer and Lehberger 2017). Upon her arrival between 9 and 10 pm, the party is already well under way. Around 60 KSK soldiers participate. The challenge in the training course has already begun. According to 'Anna's' recollections, the challenge includes a steeplechase and

[1] For further information on the KSK, see Rose 2009; Gaschke 2010.

[2] All direct quotes as well as the Facebook posts were written in German and translated by the author.

throwing pigs' heads. Once the challenge is over, 'Anna' is taken to the commander, whose hands are still covered in pig's blood. According to the script, a tent is prepared as the location for the sexual intercourse. However, sex does not take place because the commander is already too drunk.

Initially, 'Anna's' story is met with incredulous amazement among the journalists of *Panorama*'s editorial team. Nevertheless, they agree to explore the issue further. Members of *Y-Kollektiv*, a network of journalists, engage in further investigations.[3] These journalists are Dennis Leiffels, who chairs *Y-Kollektiv* as Head of Content[4], Jochen Grabler of *Radio Bremen*, and Johannes Jolmes, who is a member of *Panorama*'s editorial team (Leiffels et al. 2017). In the course of their investigations, they request information from the German Army Command in Strausberg, which refuses an interview, but replies by email and, at least partly, confirms 'Anna's' story. In addition, the email reveals more details regarding the training course, which was supposed to be a Roman or Medieval contest including "archery, slashing melons and pineapples with a sword", and "throwing pigs' heads" (Grabler et al. 2017, 04:17–04:27). In addition, log cutting with an axe and a military climbing wall were reported to have been part of the training course (*bundeswehr-journal* 2017). The internal investigation by the Bundeswehr further concludes that sex between 'Anna' and the commander did not take place.

However, the training course was not the only thing. 'Anna' reports that the crowd gave a number of Hitler salutes, sometimes even after explicit encouragement by the commander. She also states that the participants sang along to music, the lyrics of which she remembered as deploring the 'murder' of Germany. The next day, she identified those lyrics as coming from songs by right-wing rock band *Sturmwehr*. Dr Matthias Quent, head of the Institute for Democracy and Civil Society in Jena, who is interviewed as an expert in the documentary, explains that these lyrics refer to the demise of Nazi Germany and are thus clearly extreme right-wing.

Yet, the Bundeswehr's investigation does not confirm 'Anna's' information regarding the Hitler salutes and the extreme right-wing music. According to their interviews with participating KSK soldiers, the salutes were not meant to be Hitler salutes, but rather 'Ave Caesar' gestures (Gebauer and Lehberger 2017). The Bundeswehr further states that the interviews do not substantiate the suspicion of any illegal misbehaviour by KSK soldiers. Nevertheless, the "hiring of an escort woman" may indeed constitute a "violation of the soldierly obligation to integrity" (Gebauer and Lehberger 2017). Shortly after the incident was made public, a press officer of the public prosecutor's office in Tübingen announced that an investigation into the incident would be initiated (Gebauer and Lehberger 2017). Some time later, the public prosecutor's office in Stuttgart was reported to have launched an investigation based on the suspicion of symbols of unconstitutional organisations having been displayed (Gebauer 2017).

[3] For further information, see http://presse.funk.net/format/y-kollektiv/

[4] See http://tincon.org/speaker/dennis-leiffels/

The journalists of *Panorama* also see strong evidence for extreme right-wing behaviour among KSK soldiers. The soldier who established contact with 'Anna' indirectly confirms her story on the Hitler salute, because in later communication with her he declares that he "does not agree with Hitler salutes" and that he "did not know how the wind blows" (Grabler et al. 2017, 07:53–07:57). Moreover, according to the investigation by *Y-Kollektiv*, the commander in question has a large tattoo that depicts the Chetnik flag with the Serbian inscription 'for king and fatherland; freedom or death'. This flag is reported to have been very popular among radical Serbs involved in the Balkan war and the massacres of Bosnian Muslims (Leiffels et al. 2017, 10:37–11:05). Furthermore, there are doubts at *Y-Kollektiv* that elite soldiers – who adhere to a strong esprit de corps and who are most likely trained to endure severe interrogation when captured as prisoners of war – would confess their guilt during the Bundeswehr's interrogations. Indeed, by "declaring themselves to be guilty of having given the Hitler salute or to have listened and sung along to extreme right-wing music", they would "only be harming themselves as well as the KSK" (Leiffels et al. 2017, 12:05–12:18). In addition, the day after the feature was broadcast, it was reported that, based upon a complaint by a female civilian employee of the KSK, an investigation against deputy commander Colonel Thomas B. was launched for using sexist and misogynist language and threatening the female employee following his relocation (Gebauer 2017).

'Anna's' story raises the question whether there are problems of misbehaviour and extreme right-wing tendencies in the Bundeswehr in general and in the Special Forces in particular (Gessenharter et al. 1978; Wiesendahl 1998). The ZEIT Online (2017) thus comments that "in recent months, several incidents" in the Bundeswehr had "become known to the public where extreme right-wing tendencies, mobbing or sexual harassment played a role." The comment piece explicitly refers to the scandal about the allegedly extreme right-wing Bundeswehr soldier Franco A., who was suspected to have planned an attack, as well as to the media reports on the incidents of sexual harassment in Pfullendorf and Bad Reichenhall (ZEIT Online 2017). The magazine also underscores that "charges of right-wing extremism against the KSK have already existed for quite some time", pointing out retired Brigadier General Reinhard Günzel, a former KSK commander who was relieved of his office in 2003 by the then Minister of Defence, Peter Struck (SPD), "after publicly applauding a speech by former CDU member Martin Hohmann that was broadly criticised as being openly antisemitic" (ZEIT Online 2017). Günzel attracted media coverage on several occasions in the following years. In the volume of photographs on the KSK, *Geheime Krieger*[5], he explicitly puts the KSK in one line of tradition with the German command troops of World War Two called "*Brandenburgers*", a dubious special unit of the Wehrmacht (Günzel et al. 2006; Teidelbaum 2008, 11). Based on this, Lieutenant Colonol Hans-Günther Fröhling (2008) wrote in the Bundeswehr magazine *IF – Zeitschrift für Innere Führung*: "Since Günzel had been head of the KSK for three years, his statements expose the KSK to the suspicion that, at least

[5] Eng. Secret Warriors.

under his leadership, members of the KSK were perhaps socialised in a spirit that runs against the principles of *Innere Führung*."

In the following, the objective is not so much to answer the question of the extent to which the charges of right-wing extremism against the KSK are substantiated. Rather, the focus is on the public's responses to the feature by *Panorama*, based on the analysis of the thread on *Panorama*'s Facebook page.

3 Analysis of the Thread on *Panorama's* Facebook Page

Viewers' responses to the *Panorama* feature as displayed on the magazine's Facebook page were quite numerous compared to the responses to other features in the same issue. The feature itself and a condensed clip of it received more than 224,000 clicks combined, and were commented upon 304 times within just a couple of days. By comparison, the subsequent three features received just 85,000 clicks, and the feature on judges under stress in the same issue only received 44 comments. The sample analysed for this contribution consists of the abovementioned 304 comments on the KSK feature, the last of which were posted on 23 August 2017.

With regard to the viewers' posts it has to be acknowledged that it is sometimes hard to identify whether the name of the author of a post is a real name or an alias. While only a few posts openly reveal that the user was – or still is – a soldier of the Bundeswehr, with other posts it is still quite likely that the user does, in fact, have a military background. The majority of posts, however, do not allow for a clear identification of a military or non-military background. Nevertheless, this does not inhibit an inductive content analysis according to Mayring (2015). By way of such an analysis of the empirical material, the posts were examined in terms of the themes they entail. These themes were then synthesised into thematic fields. Following this procedure, the posts could be categorised along the following five thematic fields: (1) questioning the story, (2) respect and leniency, (3) shock and concern, (4) the Minister of Defence and her discharge of office, and (5) tradition and identity.

3.1 Questioning the Story

A rather large number of posts reveals substantial doubts as to whether the story that was reported actually resembles the truth. "This is the most unbelievable crap I have read in a long time", writes Jan Hammel.[6] The doubts start with the witness herself, whose credibility is questioned very much. Arne Rosenow, for instance, points out that although "everyone's talking about Nazis and Hitler salutes", it has "not been

---

[6]All comments in the following chapters were posted on *Panorama's* Facebook page, see Panorama 2017.

proven yet that this is what happened. There is a statement. But who's to say that this lady did not make up parts of her story?" This question is raised increasingly, the more viewers perceive the behaviour of the witness as morally reprehensible. Anja Baffour states rather diplomatically that she "can really only wonder about this young lady." On the same day, Dagobert Engelhardi expresses his amazement with regard to 'Anna's voluntary participation in this farewell party, describing it as "very, very odd behaviour." Margith Turner is more explicit by asking herself, as well as the readers of her post, "which decent woman would agree to serve as a sex present." For a large number of users, 'Anna's' story is nothing but implausible, and is interpreted as a reaction to her unfulfilled expectations. In Sven Hoerl's opinion, it all seems "somewhat strange. But perhaps it was the frustration of being left unsatisfied. How one could willingly agree to be presented as a sex toy at such a party remains completely incomprehensible to me." Arian Richter agrees by asking what kind of woman Anna is, seeing as "she just came to be 'taken roughly' by the commander. As he was too drunk to satisfy her, she reported Nazi slogans. Women like her are worse than the soldiers showing national pride." For Uwe Vetter it is clear that Anna did what she did because she was "disappointed that he was too drunk." Holger Rogge's comment, finally, is the most explicit:

> A woman known for liking to be fucked for free by tough boys is invited by a KSK soldier to a farewell party for his commander to fulfil his sexual desires. The commander is too drunk to fuck the horny lady. This frustrated woman 'opens up' to a team of journalists who are very critical of the Bundeswehr. The result is a story based solely on the statement of the abovementioned woman. For me, this is not credible (Panorama 2017).

The doubts about 'Anna' evoke profound scepticism regarding the feature's validity and the information revealed, which leads to strong reservations against the quality of the underlying journalistic work. Diemut Schmidt writes that "the media hype surrounding this event is highly dubious and simply a consequence of the silly season. … God knows there are enough real problems. Is it really necessary to inflate such rumours?" Markus Gebauer, in turn, more specifically refers to a "bad feature, re-enacted scenes, cheap comic images. It's not fake news, it's fucking news. And I'm supposed to get my view of world events from such nonsense?" In his post, Tobias Ruppert summarises the situation as follows:

> Well, there was this training course, which at a private farewell party certainly does not constitute official or criminal misconduct. And the young lady's statement is more than questionable, to say the least, as far as the part with extreme right-wing rock music (tasteless, but not prohibited) and the raising of the right arm is concerned. What remains is a poorly researched feature made for the common outraged citizen. And I pay for such crap with my media fees? (Panorama 2017).

Margith Turner also complains about "sensationalist journalism" and deplores being "forced by the state to finance such a farce." Tine Jänschke ironically speaks of "top journalism" and congratulates the team of *Panorama* by saying how proud she is of them. Hans-Jürgen Lemke is also concerned about the quality of the journalists' work, stating that *Panorama* was simply "puffing up random trash.

Everything that was reported is just nonsense." In line with this, *Panorama* is requested to report in a sober, objective, and cautious way. Arne Rosenow, for instance, appeals to the presumption of innocence, which "should also be clear for the media." Similarly, Thomas Johannes Schmidt wonders what actually happened:

> Tough guys who are expected to risk their lives are having a farewell party. And they don't sing 'Praise the Lord', but throw pigs' heads. It's a scandal. They invite a woman who knows exactly what she's getting into. Scandalous! Yikes! Well, nothing happened, the guy was too drunk. And now – this is what remains of the spectacular story: They listened to extreme right-wing music, perhaps even sang along, and gave the Hitler salute. Excuse me, but given that the story has been in the media all day, it's really thin (Panorama 2017).

Others go one step further, calling into question the journalistic integrity of *Panorama* and accusing the team of tendentious reporting with political inclinations. René Beer formulates his criticism as follows: "What kind of stuff do you report here? With *Panorama* you know what you get: leftist news culture directed against patriots. (…) A smear campaign against patriotic citizens serving our country. It's a shame how far we've come." Markus Köckert agrees that just because…

> …soldiers presumably listened to the 'wrong' kind of music and might have raised the wrong arm while drunk, a colourfully illustrated special feature is broadcast here? I have seen the feature, and in particular the cynical comments by [the female journalist] Reschke clearly gave me the impression this was just blind propaganda against our armed forces (Panorama 2017).

Finally, Carmen Knut points out that the "investigation isn't over yet, and you've already identified the culprits. The woman was probably disappointed that she wasn't… and then comes up with this story. What did you pay her? That's the question that needs to be asked here." She further states that she deems it "an impertinence to drag our elite soldiers into the mire. You can't even imagine what they have to go through."

3.2 Respect and Leniency

The last comment points to the second thematic field – the respect and appreciation for the job that KSK soldiers do in service of country and society. Holger Seiz, for example, insists that he is "proud of these men and their service to Germany." Mark Us Vel points out…

> …that the KSK is not a command that plays in the sandbox in case of emergency! These men face life-threatening situations every day in training, let alone in the line of duty! For what? FOR GERMANY! They are defending our country! (Panorama 2017).

This indicates a certain tolerance and leniency when it comes to behaviour that is not completely in line with the official rules of behaviour for KSK soldiers. In the same vein, Max Sunwell states that one "can't win a war with computer geeks and mamma's boys!!! ... Leave the boys alone!!! Ungrateful bastards!!!" Ines Brumm also expresses her understanding by saying that the way in which people...

> ...get off on this shit, [is] totally ridiculous. Let the boys party as they wish. Sad enough that the Bundeswehr will soon consist only of wimps wearing pink ruffles who will never ever be able to protect their homeland at the risk of their lives. Homeland – is one allowed to say that, or is it also a relic of the Nazi era? (Panorama 2017).

Shortly thereafter, Wolfgang Hansert comments with strong certainty that even "if this was the case: they are young men. They are not Nazis. This is just national pride in memory of our soldiers. If they die for Germany during foreign deployment, nobody gets upset, you hypocrites." Roland Gillig agrees, stating that...

> ...if the party's motto was Roman-Medieval, the praetorian salute is also part of it ... personally, I don't mind the fact that patriotism and even nationalism are apparently widespread within the [Bundeswehr] (amazing – I would have thought that extreme left-wing German haters or pacifist neo-'69ers would enlist, but it rather seems to be people with national pride) (Panorama 2017).

Bert Grönheim points out the extremely dangerous situations soldiers face and calls for leniency: "These are soldiers who carry out operations with a high risk of death or injury. They should be left to party in peace." The same opinion is most explicitly formulated by Ralph Bauer, who states that the KSK is...

> ...a command that is constantly at the edge of all human abilities, training and testing limits in terms of mentally extreme situations to be able to professionally deal with absolute horror scenarios in case of emergency. All methods of killing and survival in extreme situation must be mastered. This is not the Christian, moralistically and humanistically oriented bunch Ms von der Leyen would prefer. For killing machines and people who have to kill in order to ensure the survival of their compatriots, everyday humour and sarcasm looks a little different compared to the average CDU Bible group and local association members. Someone who might have served in Afghanistan, Syria, Mali [or] Kosovo sees life and death and everything that goes with it more realistically than some scribbler who judges solely based on theory. War tactics, war psychology etc., and the training they entail are always on the edge of what is normal. Therefore: when such state-paid killers and killing machines have a party, it's just rough. ... Ms von der Leyen presides over troops who must love their country to such an extent that they kill and be killed for a few Euros, and who are supposed to show absolute obedience, unconditionally for the homeland, but in a Christian, kind, friendly, humanistic and moral way without any complaint, if you please. My dear friends, such soldiers don't exist. They especially don't exist in special killing squads. ... If you don't want such stories and pictures, stop your wars and get rid of armies and soldiers (Panorama 2017).

Moreover, one may not even speak of a real scandal. "Hand on heart: not nice, but no scandal either", writes Anja Baffour. This opinion even seems to be reinforced when taking into consideration that alcohol was involved. "A little fun has to be allowed. That was certainly no Hitler salute. The guys were in party mood and drunk", states Dagobert Engelhardi. Similarly, Heinz Hellbach comments that the only thing he finds "reprehensible is throwing pigs' heads", further stating that he has "sympathy for all those, including our police, who deal with human scum every day. That they also need their fun sometimes to wash away all the pent-up frustration with binge drinking. Does it really matter if they behave a bit badly when they're drunk?" Kai Ullrich, in turn, writes that he would...

> ...also prefer to be defended by leftist, second-chance education social pedagogues! There are some things that one just has to accept. ... These are extremely highly trained contract killers, for Christ's sake! There is no transgender debate, nor should some kitchen maid go public (Panorama 2017).

Others point to similar incidents that did not have such consequences. Thus, Ulrich Kesse states:

> Don't really understand what all the fuzz is about. I was with a combat unit (tank grenadier in 1978/80. During platoon or company parties after military manoeuvres, it was not unusual that once in a while one or the other 'forbidden' song was sung or that the Hitler salute was given. Let these guys live. After all, they risk their lives for our country. I for my part can fully understand them (Panorama 2017).

Finally, Alexander May in this context refers to the democratic right to freedom of speech:

> We must get back to acknowledging that what many might find tasteless is nevertheless covered by the right to freedom of expression. Even giving a Hitler salute or the indefinable so-called hate speech should be covered by freedom of expression. Just like, of course, the counter-speech to it (Panorama 2017).

3.3 Shock and Concern

Other commentators, however, show little surprise and do not dispute the authenticity of the story. Martin Armbrust, for instance, is "not surprised that the KSK are involved in such activities. They don't know what they stand for: a democratic society that trusts in human rights. I do not have the slightest doubt that these accusations are true." Tom Meyers agrees, stating that the "military attracts right-wingers like a dunghill attracts flies." Jonas Hortebusch ironically asks: "Really? Nazis in the Bundeswehr? I am shocked! Now don't tell me that Elvis is dead too! That would be too much 'news' for one day!" In the same vein, Tobi He comments: "Are you kidding? Nazis in the military? Nobody could have expected that in such a social and not at all reactionary association." In this regard, other commentators point to similar incidents in the past, presuming that they are the result of a structural problem. Niels Petring, for instance, is genuinely confident that it "is the military system, and these are not isolated cases." Elio Bauer agrees:

> 20 years of individual cases? Since the end of the World War II we've been cuddling up to Nazis. The past was never really processed, nor was there a denazification, whether in the military, the justice system, politics or the business world (Panorama 2017).

For some, the report was simply the tip of the iceberg. "What would remain of the 'command' if all the Nazis were removed? Empty barracks probably", writes Matthias Wetzel. Others point out the suspension of the draft. Josef Theobald, for instance, interprets the incident as...

> …the dark side of the volunteer and professional army. Only, this was politically intended. With a conscript and professional army, this could have been avoided due to a different socio-cultural composition of the armed forces. Now we must more often act rigorously. Since new recruits are often only to be found in extreme right-wing circles, it will be difficult to recruit the right people (Panorama 2017).

Red Mask agrees and wonders why people are even surprised:

> There "have always been Nazis in the Bundeswehr. If you abolish conscription, it is more than 'normal' that only people with a strong love to their homeland feel called to the military. In my time, back in 1990, the fascists in the company were classified as oddballs …. At that time, none of the Nazis dared to openly confess their political convictions (with Hitler salutes or memorabilia) – today … the 'national' is lived out unrestrainedly. You always get what you create for yourself (Panorama 2017).

Bernd Adrian agrees that "things are likely to get even worse" because, since conscription has been abolished, the Bundeswehr now attracts even more "gun freaks as well as right-wing extremists." In addition, commentators point to mistakes and omissions in the aptitude test, which may have contributed to the deplorable present situation. In this regard, Maxe Baumann argues that the aptitude tests should take more efforts to identify recruits who have a "morally consolidated world view", and should not only focus on "how low their inhibition threshold for killing is." Moreover, according to Matthias Wetzel, a certain co-responsibility should be attributed to the superiors, who are deemed to have known "whom they were training to be killers." Christa Elli Schonscheck agrees that "something like this does not happen overnight. Such processes have a start-up period in which there were obviously never any controls." It could be possible, as Daniel Lücking suspects, that such attitudes are particularly pronounced in the fighting forces, since…

> …especially among the fighting troops … 'traditionalists' are to be found… Personally, I'm not surprised that now even officers are involved. The ideology has come to light again and again in the last decades. There is little that can be done about it – especially in the KSK, which are very committed to one another and form a closed group from which no one can be separated. No subordinate will betray the boss or report someone else (Panorama 2017).

While some of the posts show a kind of succinct cynicism, others fall prey to fatalism. Konrad Struck, for instance, is certain that "the Bundeswehr will never get a hand on this" and does not even "want to know what's being swept under the carpet." Andreas Reymann nourishes a similar scepticism, stating that it is...

> ...hypocritical to have people believe that these are exceptions in the lower ranks. The Bundeswehr … [is] per se completely anti-democratic, and therein lies the danger of the present times, since all parties in parliament support the reorientation of German foreign policy … and want to rely increasingly on these institutions, finance them and protect them from criticism instead of controlling them (Panorama 2017).

Most of those observers who do not doubt the report see the misbehaviour of the KSK soldiers as a potential threat to the democratic political system in Germany and call for consequences. Thus writes Klaus-Dieter Kroll: "Unbelievable, and they are meant to protect and defend us?" Matthias Dittmann, in turn, sees the feature by *Panorama* as an unmistakable warning to politics, since it "should give the Federal

Government something to think about when even the Bundeswehr elite thinks completely differently than the Democrats want! "Walter Borgius is even more explicit and sees the political elites in Germany called to action:

> The fact that there are neo-Nazis in the Bundeswehr, doesn't shock me much. That is why it is all the more important to keep an eye on these circles, which are armed and politically highly dangerous (Panorama 2017).

Thus, many posts call for severe consequences for what happened. Norbert Gesser, for instance, believes that "the Bundeswehr is no place for extreme right-wing ideologies. And where they are spread illegally, those responsible should be punished." Karl Lochner agrees, stating that if "the feature shows the truth, everyone involved should be suspended from service, without any severance pay or other claims." In the same vein, Andreas Mitterhofer points to the commitment of elite soldiers to the democratic order, which he deems "simply incompatible with the Hitler salute", thus justifying the dismissal of these soldiers. Patricia Castañeda-Holmsve is fully in line with this when she argues that…

> …people with those sentiments don't belong in the army. And such medieval games simply have nothing to do with fun anymore, they are sick. Hopefully this incident will have consequences (Panorama 2017).

Many of the commentators posting these opinions also do not understand why many posts display substantial and far-reaching understanding for the elite soldiers' misbehaviour. Mike Zeh, for instance, believes that the…

> …witness seems credible in any case … Those who now claim that soldiers are allowed to utter such slogans outside of duty time have misunderstood the responsibility of a state authority. Why, they have an exemplary function, also towards future generations. A Hitler salute has not a place here, not even in private. One should be allowed to expect this much discipline, after all, they are also paid for by tax money (Panorama 2017).

As a consequence, Axel Wellinghausen insists that the incident be…

> …investigated very thoroughly. Above all, investigators should be allowed to do their job, not hindered. Should they prove disciplinary and/or criminal misconduct, it should be punished as a last resort (Panorama 2017).

Indeed, some of the posts in this thematic field specifically comment on the journalistic work that was involved in producing the feature, and which is seen as part of the investigation. Walter Borgius thanks *Panorama* for the contribution, stating that the "hate and nonsense here in the comments should only encourage" the team to "continue with their critical journalism. … Thanks a lot for your work."

3.4 The Minister of Defence and Her Discharge of Office

On the eve of 30 April 2017, the ZDF programme *Berlin direkt* broadcast an interview with the Minister of Defence, Dr Ursula von der Leyen, which focused on the scandal concerning the allegedly extreme right-wing soldier Franco A. In the interview, the Minister put the case of Franco A. within a larger framework of further incidents and talked about a leadership failure by superiors, stating that…

> …these are all different cases, but for me they now belong to the same pattern, so that today I say: the Bundeswehr has an attitude problem. And it obviously has a weakness in leadership at various levels. And we must take a consistent approach to deal with this (Panorama 2017).

Based on this statement, it is not surprising that the posts commenting on the Panorama feature also discuss the Minister of Defence, especially after she had officially characterised the featured incident as "absolutely tasteless" on 22 August 2017. In this regard, she receives both consent and thorough criticism. Hans-Jürgen Lemke sees the Minister's actions as coming from an unsubstantiated, excessive desire for action, which is leaving him…

> …quite fed up. Why can't we just give it a rest? Since … Uschi has been stirring things up, the troops consist of Nazis only. Then comes *Panorama* and also makes a big deal out of it. What was reported there is complete nonsense. I was in the army for a long time, and I have never ever experienced anything like this (Panorama 2017).

Jens Simon agrees that the Minister can "finally dismiss a commander, as well as all the soldiers involved. Again, the Nazi witch's hammer can be taken out, and the Berlin inquisition will hunt and publicly burn them."

In addition, the Minister is criticised for what is interpreted as insufficient loyalty towards the troops she commands. Roland Fleischmann, for instance, states that one "may well find this party … tasteless. One can also criticise it *internally*. But one has to be loyal to one's people, and I miss that very much when it comes to the first female Minister of Defence Germany ever had." Frank Rosemeyer shares this view, voicing his indignancy that the Minister, who "is the one in charge", could be "insulting her subordinates" thus.

Others use the Minister's statements to unleash thorough criticism on her job performance. Heinz Scharrer, for example, uses his post to request from the Minister that she should "first learn how to do" her job. Moreover, Nils Lessmann believes that, as a consequence of the Minister's job performance, the Bundeswehr has become "nothing but an underfinanced, castrated, and gendered bunch of people." Both the integration of women into the Bundeswehr and the positive response to societal demands are seen as a softening of the armed forces, with negative repercussions on military readiness and military effectiveness. In this regard, Thomas Müller hopes to "be rid of this Defence Minister actress soon." Joe WT, finally, states that, since the Minister of Defence "does not seem to have her business under control, she should clear out and make room for someone who does."

However, there is also quite a lot of support for the Minister. Marcel Heldt, for instance, thinks that "one can nothing but praise this woman." Similarly, Florian Friedrich is happy to have someone in charge of the Ministry of Defence "who is not enmeshed in the military's cadre mentality." In much the same way, Ingrid Rausch wonders whether the Minister should really be expected to "be loyal to the extreme right-wingers in the Bundeswehr." On the contrary, it is, as Maria Stein argues, necessary "to muck out the Bundeswehr." Moreover, Matthias Schlott calls for "disciplinary consequences", which, in Christa Elli Schonscheck's opinion, must also apply to the superiors, who apparently had not been conducting any controls. Manfred Willi Reichert does not see the necessity of the Minister to be loyal to her troops either: "Why does one have to be loyal, and how far should loyalty go when it comes to breaking the law?" Leonardo Cucchiara is the most explicit when pointing out that the Minister had no choice but to "criticise such actions":

> Precisely the greatest obstacle to investigating crimes within the Bundeswehr is the fact that silence is dictated from above because they want to 'settle it amongst themselves'. There is definitely a widespread problem with latent nationalism in the Bundeswehr. … Everybody knows about this ghost in the troops. … And it is clearly a problem of leadership. Unfortunately, only a few people dare to speak up when they see something like this. But even if it comes out somehow, it is completely played down. Even by the upper echelons. Unfortunately, I experienced this myself. … The way I see it, the Minister is quite right. And she could be even tougher if she wanted to (Panorama 2017).

3.5 *Tradition and Identity*

Against the background of a new debate about traditions in the Bundeswehr and the revision of the edict on traditions it is not surprising that this thematic field is also covered in the thread. Andreas Reymann finds it "disgusting, but unfortunately not surprising, that the most undemocratic institution in the state cultivates traditions of the Nazi regime." In contrast, Nils Lessmann, who wrote several posts on this topic, ironically characterises as bad the fact that there are "soldiers who are proud of their country and cherish traditions. Would you rather like to see that leftist anarchists serve in the Bundeswehr?" Shortly thereafter, he adds that the "Bundeswehr is the successor to the armed forces of the previous German states (German Empire, Weimar Republic, Wehrmacht) and should therefore preserve these traditions." Finally, he argues that "a strong military needs patriotic soldiers. Powerful armed forces need a strong esprit de corps, which is consolidated by military traditions."

Christoph Habereder, on the other hand, objects to this stance, arguing that "maintaining tradition can become a problem and certainly requires special sensitivity because of our past." In this context, Claus Bier warns that "patriotism can easily be confused with nationalism", and that it is thus all the more justified to ask "why the tradition of an army that has lost two (offensive) world wars should be upheld. Why has the Bundeswehr not managed to develop traditions of its own?" Furthermore, while "a certain healthy patriotism in the military" is necessary, Sajeel Ahmad believes that with regard to this incident the boundary to racism has been crossed. He points out that "Nazi symbols, the Hitler salute, listening to misanthropic music are not patriotism, but pure racism. There is indeed a difference between patriotism and racism." Christopher Müller, in turn, does not agree that patriotism is necessary because the "only right and important patriotism that should prevail in the Bundeswehr is constitutional patriotism …. Extreme right-wing ideologies are unconstitutional!"

However, there may be even more at stake, as Norbert Grünewald argues. He places traditions within the larger context of identity, believing that "the Bundeswehr has a problem with identity." Finally, Oda DeVito puts the issue in an even more complex frame, arguing it to be a problem of society at large:

> Reading the contributions in this thread, it's hard to know which is scarier: the filthy insults and the defamation of the editorial team, the journalists as complete fools and left-wing hacks who allegedly have no idea of serious research and professional validity? Or the glorifying subservient spirit which trivialises the alleged incidents in this 'elite unit' as harmless celebratory excesses? Or the blatant expressions of sympathy for these 'boys' who allegedly 'risk their arses' every day to protect us citizens, while the fee-financed media blows up any uninteresting crap out of sensationalism? Or the sexist and disgusting insults and humiliations of these self-righteous defenders of the troops, who slander the witness as a slut, whore or horny informer? One thing is clear: We do not just have a problem in the armed forces, we have a far greater problem in our society regarding our collective values (Panorama 2017).

4 Conclusion

Analysing the posts in the *Panorama*-thread on Facebook belongs to the comparatively young research area of "social media analytics" (Stieglitz et al. 2014). In social media, a small group presents itself to the public and to public opinion. In order to gain access to this area, one needs to have access to modern information technology as well as know how to use it. The public of social media is thus not representative of the public at large. Nevertheless, the analysis of such a partial public may be relevant to the overall societal discourse.

In the case presented here, the inductive content analysis of the thread that developed from the *Panorama* feature revealed and identified five thematic fields covered in the posts.

1. The posts in the first thematic field nourish deep concern and doubts regarding the truthfulness of the feature concerning the events that occurred at the KSK farewell party.
2. The posts in the second thematic field call for respect for the professional performance of the members of this KSK unit, showing considerable leniency concerning the alleged or actual misbehaviour of the elite soldiers.
3. The third thematic field focuses on the shock regarding the reported incident and the ensuing concern about its impact upon the functioning of democracy in Germany.
4. The fourth thematic field is represented by posts which take the feature as a welcome starting point to comment on the person of the Minister of Defence and her discharge of office.
5. The posts in the fifth thematic field, finally, consider the incident within the much larger context of military traditions and military identity.

The debate within this partial public represented on *Panorama*'s Facebook page does not have losers and winners because neither side can undisputedly claim victory in the debate. The positive aspect, on the one hand, is the fact that the 'checks and balances' are working – the advocates for leadership development and civic education (Innere Führung) participate in this debate and articulate their position. On the other hand, the negative aspect is the fact that one has to acknowledge that the contents and the principles of leadership development and civic education are by no means undisputedly shared by this partial public. Many posts can be read as a criticism of leadership development and civic education, which proves this criticism is also shared in parts of the general society. The fight for leadership development and civic education is by no means over – it rages on both within the military and in the general public.

As regards the Bundeswehr's relationship to social media, the present case is interesting since the Bundeswehr as an institution is not – at least not openly (in the world of today nothing should be ruled out) – an actor in the discourse on *Panorama*'s Facebook page. However, the Bundeswehr is the focus of 224,000 clicks and more than 300 comments by members of the interested public. These comments come from civilians as well as from former or present members of the armed forces. They are individual responses and reactions to the documentary on the alleged incident in the German Special Forces. The Bundeswehr, then, is talked about on social media whether it wants to or not, and there is little that can be done about this. Consequently, this means that controlling social media is more or less impossible. The Bundeswehr may, of course, engage in the dispute officially as an institution, although this might lend more importance and political meaning to the thread than would be deemed appropriate by the organisation. Thus, it may be hoped in the public relations offices of the Ministry of Defence that this thread vanishes, so to speak, and is lost in the maelstrom of the internet and the news overload that comes with it. However, compared to the past, the debate is no longer simply verbal, but is held in written form and as such stays on the internet forever. As a consequence, although the debate on this incident may well be buried under tons and tons of other news, it can still be found very easily – all it takes is a search command in a search engine such as Google and others. The incident and the debate on it can be retrieved again and again, even after considerable time has passed. Welcome to the Brave New World of social media.

References

Berlin direkt (2017) Personalprobleme in der Truppe: Interview with the Minister of Defence, Ursula von der Leyen [online]. Berlin direkt. [Viewed 29 August 2017]. Available from: https://www.zdf.de/politik/berlin-direkt/berlin-direkt-vom-30-april-2017-100.html

bundeswehr-journal (2017) Bei Abschiedsfeier der KSK "die Sau rausgelassen": Recherchen zu KSK-Feier. Bundeswehr ermittelt bei Elitetruppe [online]. Bundeswehr-journal. [Viewed 29 August 2017]. Available from: http://www.bundeswehr-journal.de/2017/bei-abschiedsfeier-der-ksk-die-sau-rausgelassen/

Fröhling HG (2008) Mehr Transparenz [online]. IF – Zeitschrift für Innere Führung, 1. [Viewed 29 August 2017]. Available from: http://www.if-zeitschrift.de/portal/a/ifz/start/themen/buerger_staat/!ut/p/z1/hY_NCsIwEITfqJukaOvRYivFWhV_k4uEJtRITUqI4sGHN0Hx-VtzDwM7sfssCgxMwzR-q5U4ZzTvfUzY-Z2m1q8iEkBneEFTuyzSLlzuS5zEc4Phvh-PkYDdQUwVZIoJ6RDDL2Y9gCAyZk1BgtXVAntVNeW8udsVFvrOtCcrfWJ5ESQ-BGeZTjG5HsKv5KCLubzBOGyLlYBeOUP_vzt8iY8DfTCtejk2jTTj9HfirSuR-0bjVwyPA!!/dz/d5/L2dBISEvZ0FBIS9nQSEh/#Z7_B8LTL2922D1Q20IUI8B3MT2EU6

Gaschke S (2010) Unter Kriegern: Sind Elitesoldaten auf geheimer Mission die Zukunft der Bundeswehr? Ein Besuch beim Kommando Spezialkräfte [online]. ZEIT-Online. [Viewed 29 August 2017]. Available from: http://www.zeit.de/2010/31/KSK-Kommando-SPezialkraefte/komplettansicht

Gebauer M (2017) Ärger in Elite-Einheit. Frauenfeindliche Sprüche: Bundeswehr setzt KSK-Kommandeur ab [online]. Spiegel online. [Viewed 29 August 2017]. Available from: http://www.spiegel.de/politik/deutschland/bundeswehr-ksk-kommandeur-wegen-frauenfeindlicher-sprueche-abgesetzt-a-1163440.html

Gebauer M, Lehberger R (2017) Bundeswehreinheit KSK: Ermittlungen nach abstoßender Feier von Elitesoldaten [online]. Spiegel online. [Viewed 29 August 2017]. Available from: http://www.spiegel.de/politik/deutschland/bundeswehr-ermittelt-nach-geschmackloser-party-von-ksk-soldaten-a-1163290.html

Gessenharter W, Fröchling H, Krupp B, Nacken W (1978) Rechtsextremismus als normativpraktisches Forschungsproblem: Eine empirische Analyse der Einstellungen von studierenden Offizieren der Hochschule der Bundeswehr Hamburg sowie von militärischen und zivilen Vergleichsgruppen. Beltz, Frankfurt am Main

Grabler J, Leiffels D, Jolmes J (2017) Hitlergruß? Ermittlungen gegen Kompaniechef. Rechtsrock und Hitlergruß auf einer Verabschiedungsfeier für den Kompaniechef? Bundeswehr und Staatsanwaltschaft haben Ermittlungen aufgenommen [online]. Panorama. [Viewed 29 August 2017]. Available from: http://www.ardmediathek.de/tv/Panorama/Hitlergru%C3%9F-Ermittlungen-gegen-Kompaniec/Das-Erste/Video?bcastId=310918&documentId=45275570

Günzel R, Walther W, Wegener UK (2006) Geheime Krieger: Drei deutsche Kommandoverbände im Bild: KSK, Brandenburger, GSG 9. Pour le Mérite, Selent

Leiffels D, Grabler J, Jolmes J (2017) Ermittlungen bei Bundeswehr-Eliteeinheit: Hitlergrüße und Rechtsrock [online]. Y-Kollektiv Dokumentation. [Viewed 29 August 2017]. Available from: https://www.youtube.com/watch?v=G_oSUzT5iw8

Mayring P (2015) Qualitative Inhaltsanalyse. Grundlagen und Techniken, 12th edn. Beltz, Weinheim

Panorama (2017) [Bundeswehr: Hitlergruss beim KSK?] [Facebook]. 17 August 2017. [Viewed 23 August 2017]. Available from: https://www.facebook.com/panorama.de/videos/1095337790603890/

Rose J (2009) Ernstfall Angriffskrieg. Frieden schaffen mit aller Gewalt? Verlag Ossietzky, Hannover

Stieglitz S, Dang-Xuan L, Bruns A, Neuberger C (2014) Social Media Analytics: Ein interdisziplinärer Ansatz und seine Implikationen für die Wirtschaftsinformatik. Wirtschaftsinformatik 56(2):101–109

Teidelbaum L (2008) Braunzone Bundeswehr? Der bundesdeutsche Rechtsextremismus und die Bundeswehr. Informationsstelle Militarisierung e.V. 4, Tübingen

Wiesendahl E (1998) Rechtsextremismus in der Bundeswehr: Ein Beitrag zur Aufhellung eines tabuisierten Themas. Sicherheit und Frieden 16(4):239–246

ZEIT Online (2017) Bundeswehr. Vizekommandeur der KSK soll abgesetzt werden [online]. ZEIT-Online [Viewed 29 August 2017]. Available from: http://www.zeit.de/gesellschaft/zeitgeschehen/2017-08/bundeswehr-ksk-kommandeur-stellvertreter-absetzung?print

Sentiment of Armed Forces Social Media Accounts in the United Kingdom: An Initial Analysis of Twitter Content

Daniel Leightley, Marie-Louise Sharp, Victoria Williamson, Nicola T. Fear, and Rachael Gribble

Abstract Prior research on the United Kingdom (UK) public's perception towards the British Armed Forces often found a contradicting understanding of the military as both 'heroes' and 'victims'. In order to examine these contradictions further, this study examined public attitudes and perceptions of the British Armed Forces, using a sentiment analysis of Twitter content posted on or after 1 January 2014. Twitter is one of the largest social media platforms, with an estimated 126 million daily active users worldwide, and 17 million active users in the UK. A bespoke data collection platform was developed to identify and extract relevant tweets and replies. In total, 323,512 tweets and 17,234 replies were identified and analysed. We found that tweets related to or discussing the British Armed Forces were significantly more positive than negative, with public perceptions of the Armed Forces stable over time. We also observed that it was more likely for negative tweets to be posted late evening or early morning compared to other hours of the day. Furthermore, this study identified differences in how positive and negative tweets were discussed in relation to politicised hashtags concerning Government policy, political organisations, and mental health. This was an unexpected finding, and more research is required to understand the reasons as to why this is the case.

Keywords British Armed Forces · Social media · Mental health · Sentiment · Public perceptions · Public attitudes

---

D. Leightley (✉) · M.-L. Sharp · V. Williamson · N. T. Fear · R. Gribble
King's Centre for Military Health Research, King's College London, London, UK
e-mail: daniel.leightley@kcl.ac.uk

E. Moehlecke de Baseggio et al. (eds.), *Social Media and the Armed Forces*, Advanced Sciences and Technologies for Security Applications,
https://doi.org/10.1007/978-3-030-47511-6_9

1 Introduction

Public support for military action plays an important role in defence and foreign policy, from establishing the legitimacy of military operations to maintaining military effectiveness, justifying defence budgets, encouraging retention and recruitment, and ensuring support for veterans (Rahbek-Clemmensen et al. 2012; Gribble et al. 2015). Since the beginning of the UK's involvement in Iraq and Afghanistan nearly two decades ago, there has been widespread concern from military leaders about what the British public thought of these missions and how public perception might influence support for personnel and veterans on their return (Hines et al. 2015; Gribble et al. 2019). However, there has been, and continues to be, a lack of UK-based research on public attitudes towards the British Armed Forces from a UK perspective.

Evidence indicating differences in attitudes and perceptions between the general public and the armed forces is referred to as the 'civil-military gap'. First described in literature originating from the U.S., it refers to the cultural and demographic gap that can arise between society and members of the armed forces due to a lack of contact, shared experiences, and demographic representation (Huntington 1957; Biderman 1960). A widening gap can have implications for mutual understanding and support, defence and security policy, recruitment and retention of military personnel, and how well those leaving military service transition back into civilian society (Rahbek-Clemmensen et al. 2012). Prior research conducted in this area suggests a contradiction in the way in which the public perceive the British Armed Forces (McCartney 2011). A study using data from the British Social Attitudes survey, for instance, found high levels of public esteem and respect for the armed forces, with "more than eight out of ten people" saying they held "a 'high' or 'very high' opinion of the Armed Forces" (Gribble et al. 2012, 141). In fact, according to the study, service personnel were more respected than any other profession, including doctors, who regularly top the list of most respected professions (ibid., 143).

Differences in perceptions may stem from decreasing public contact with those who have served, or currently are serving, in the armed forces, following the shrinking of the military establishment after the Cold War and the passing of generations who experienced conscription for World Wars One and Two and the national service – which was abolished in 1963 in the UK (Strachan 2003). Furthermore, over the last 20 years there has been a marked shift in the role(s) and responsibilities of the British Armed Forces. Active combat missions in Iraq and Afghanistan have ceased and personnel are involved increasingly in non-combat operations, such as humanitarian responses, including Operation Gritrock in response to the Ebola epidemic in 2014, supporting first responders fighting the 2018 wild fires in England and Wales, and supporting residents in England during the widespread floods in 2019 and early 2020 (BBC 2018; Forces.Net 2019). However, despite such frontline, public-facing roles, research indicates that the general public do not understand the daily roles of the armed forces, given the reduction of the media focus on their current activities, which may contribute to more negative perceptions (Latter et al. 2018).

While there is a clear admiration for the British Armed Forces, this is often accompanied by misperceptions and misinformation about the impact of military service on those who served (Rahbek-Clemmensen et al. 2012; Gribble et al. 2015, 2019; Hines et al. 2015). British research shows that the public regularly overestimate the negative impacts of military service on military personnel and veterans, especially as regards mental health. The social research institute, Ipsos MORI, in collaboration with King's College London, conducted the 'Hearts and minds: Misperceptions and the military' study, an international survey on perceptions of the armed forces, for which a total of 5010 interviews were conducted across 5 countries: Australia, UK, USA, Canada, and France (Ipsos MORI 2015). The study found several misconceptions among the general population, which included overestimating the amount spent on armed forces, overestimating mental health problems, and overestimating former service personnel's impact on the justice system (ibid.). These findings are not unusual, with previous surveys showing that 91% of the British public believe it is to be expected that former service personnel have physical, emotional or mental health problems because of their service (Ashcroft 2012).

More recent online surveys indicate that such perceptions continue. For instance, a poll conducted by YouGov shows that UK veterans are perceived as inherently likely to be institutionalised, psychologically impaired or 'damaged' due to military service (Latter et al. 2018). Moreover, there is a general perception that veterans are less able to build relationships outside the armed forces. However, at the same time, the public also believe that military service develops positive attributes, such as self-discipline, loyalty, and self-reliance. Overall, these findings seem to reflect the aforementioned hero/victim dichotomy in the British public's understanding of the armed forces and their role (McCartney 2011). It also should be noted that such perceptions could have further implications regarding the public's support of military operations and thus the ability of the Government to enact foreign policy (Forster 2005; McCartney 2010).

In the UK, there is limited research exploring public perceptions of the British Armed Forces using traditional-based approaches, such as surveys or face-to-face interviews. The increasing costs of conducting large-scale quantitative surveys – and the difficulties recruiting enough participants – may prohibit future research in this area. There has been an increased reliance on online surveys, however, they remove the ability of contextualising responses and are often done remotely without any direct contact or relationship. An alternative methodology is the use of social media to monitor public opinion in response to current events in real time. Although concerns do exist for using social media, it is important to acknowledge that the use of social media can give a vocal platform to nameless individuals to promote their ideas and thoughts freely without balance, and that this could overshow and devalue other perceptions.

There is a growing use of Twitter to quantify public opinion and perception. For example, the sentiment computed from tweets and replies to tweets has been used to characterise public perceptions of eating disorders, vaccines, illness, and pain (Ashcroft 2014; Hines et al. 2015; Mahar et al. 2017). Twitter, a microblogging service provider, has an estimated 126 million daily active users, generating over 400 million tweets per day. In the UK alone, Twitter has over 17 million active users (Morgan 2001; Szayna et al. 2007). It is therefore an ideal platform for analysing public opinion of the British Armed Forces. In this study, we examined the applicability of sentiment and content analysis of Twitter, including tweets and replies posted both by members of the general public and on armed forces accounts, in order to understand public perceptions of the British Armed Forces. Thus, we intend to provide support for alternative methods of examining public perceptions in this area and to understand how social media can advance our understanding compared to prior literature.

2 Methodology

2.1 *Study Design and Data Sources*

This study was designed as a quantitative analysis of tweets and replies – including the use of hashtags – posted on Twitter by publicly available accounts relating to, or discussing the British Armed Forces. Three members of the study team[1] manually searched Twitter to identify relevant accounts. The inclusion criteria for the accounts were:

1. publicly accessible (non-private);
2. posts relating to or discussing the British Armed Forces (i.e., veterans' charity or an individual who served);
3. written in English;
4. posted on or after 1 January 2014[2];
5. have a minimum of 1000 followers at time of data extraction.

The researchers determined each account's suitability for inclusion based on the content of the tweets and replies posted by the account. Furthermore, we used the number of account followers to indicate trustworthiness and to ensure that we did not include any fake accounts ('bots') or new accounts that may lack trustworthiness.

The exclusion criteria for the accounts were:

1. no identifiable language;
2. only contained a link or image (indicative of spam tweets, also referred to as junk or spam tweets);

[1] Researchers: Dr Daniel Leightley, Dr Marie-Louise Sharp, and Dr Rachael Gribble.

[2] Prior to 1 January 2014, Twitter processed, collated, and disseminated Twitter content in a different manner, thus making a direct comparison pre and post this date unreliable.

3. a re-tweet with no comment;
4. contain 15 characters or less.

For data collection, we used the Twitter Streaming Application Programming Interface (API), which has rate-limits to ensure fair use, meaning that only 2500 tweets can be collected per account (Twitter 2019). For each Twitter account meeting the inclusion criteria, the API was used to extract the most recent 2500 tweets and, where possible, the replies to each tweet. The tweets were automatically collected and stored in a password-protected database. Data extraction was performed in May 2019 and included tweet/reply content, coordinates (latitude/longitude) of the tweet/reply, date of creation, predicted language of the tweet/reply, method of posting, and, if there was a reply, the user at whom the reply was directed.

Figure 1 shows a flowchart illustrating the process we followed for the analysis of the tweets and replies, along with the numbers of the included and excluded tweets and replies.

2.2 *Data Cleaning*

In order to analyse the extracted Twitter content, it was important to process the tweets and replies prior to the analysis and modelling to ensure that we did not model irrelevant or non-important features. This is a standard Natural Language Processing methodological procedure used in a number of studies (Jinwei Cao et al. n.d.; Gundlapalli et al. 2013; Leightley et al. 2019). For this purpose, the following text processing steps were applied to the tweets and replies to remove tweets and replies that were not of interest:

1. removing special characters that did not provide useful information, e.g., (&), (“, “), (∗), (+), (<), (>);
2. identifying replies and mentions to other users (represented by @) and removing hyperlinks;
3. removing the hashtag symbol (#) and divide any hashtags into multiple words with the objective of reducing the hashtag to core terms.

2.3 *Sentiment Analysis*

In classical content analysis, human readers qualitatively identify themes and concepts in the text, which is time-consuming and limited. In this study, in order to estimate the sentiments expressed in each tweet and reply, we used an automated, computer-based lexicon and a rule-based sentiment classifier called Valence Aware Dictionary for Sentiment Reasoning, or VADER for short (Gilbert and Hutto 2014). VADER, which classifies sentiment as well as the valence for each word to compute a positive, a negative, and a neutral score for each sentence, has been used in

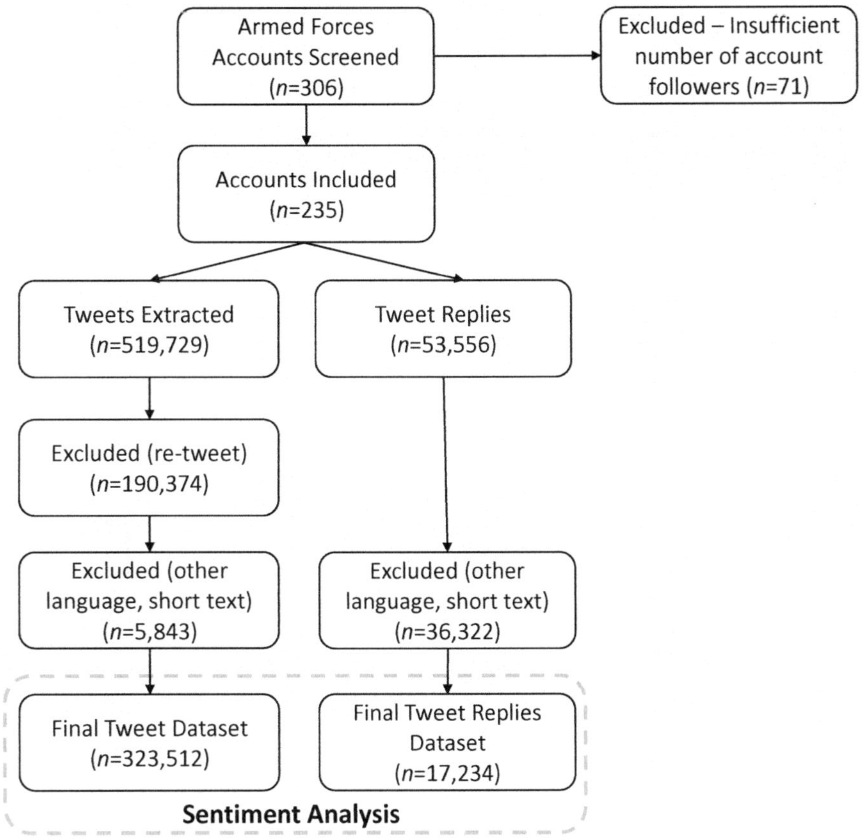

Fig. 1 Flowchart of data extraction and processing from retrieved tweets and replies

numerous studies to assess and predict sentiment (e.g., Daniulaityte et al. 2016; Ghani et al. 2018; Sewalk et al. 2018). VADER was selected as it was found to be successful in reliably handling social media text (Huang et al. 2018; Sewalk et al. 2018; Pérez-Pérez et al. 2019). We used VADER to compute the sentiment score for each sentence in the selected tweets and replies. As is practice in other approaches, we considered a sentence positive if the mean score was $\geq+0.3$, or negative if the score was ≤-0.3. The mean of all nonzero scores was then used to calculate a sentiment score per tweet. Mean scores between -0.3 and $+0.3$ were considered neutral and were excluded from this analysis (Gilbert and Hutto 2014; Sewalk et al. 2018; Pérez-Pérez et al. 2019).

2.4 Additional Data Analysis

In this study, we leveraged our analysis to evaluate not only the expressed sentiment of each tweet and reply, but also the content and metadata provided with each tweet and reply. The literature indicates that time of posting, hashtags, and geolocation – where used or collected by Twitter – can offer unique insights that otherwise would not be captured by sentiment analysis alone (Kolliakou et al. 2016; Radzikowski et al. 2016; Roland et al. 2017). Specifically, the following features were extracted and analysed:

1. The day of the week and the time the content was posted were extracted for each tweet and reply. Previous research shows strong correlations between the sentiment score and the day of the week and the time the content was posted. For example, McIver et al. (2015) found that tweets posted early in the morning were often more negative compared to those posted in the evening, which may reflect psychological state changes throughout the day.
2. Twitter allows for the use of 'hashtags' in tweets and replies, which is included in the metadata. Hashtags (denoted with the # symbol) mark keywords or phrases and are categorised by Twitter. They summarise the overall narrative of a tweet/reply and reflect which topics the poster considers relevant and important (Radzikowski et al. 2016).
3. Some Twitter users have given permission for their location at the time of posting to be made publicly available. This could prove useful in identifying their current country of origin as well as distinguishing specific country issues, aiding in further decoding and understanding of the narrative of the tweet or reply, and validating any claims made (Radzikowski et al. 2016).

It is important to note that not all Twitter users will have used hashtags or have given permission for their location to be accessed and shared.

2.5 Statistical Analysis

The analysis conducted in this study largely follows analyses described previously, and uses the programming language Python (Kolliakou et al. 2016; Sewalk et al. 2018; Pérez-Pérez et al. 2019). Firstly, we provide unweighted descriptive statistics on the frequency, standard deviation (SD), and length of tweets and replies as well as average word length per tweet/reply, post rate per account, number of likes/retweets, average number of hashtags, and average number of tweets/replies posted in the morning or afternoon. Secondly, we sought to identify whether there were any differences between positive and negative sentiment scores, using a Mann-Whitney nonparametric test to determine whether sentiment statistically changed over time. Thirdly, we identified the most popular hashtags used for tweets and replies separately, based on a frequency count. The number of positive tweets for each hashtag was compared using Chi-square statistics, which was repeated for the replies. Finally, for those tweets and replies with a geolocation marker, we grouped and identified the most popular countries in which tweets/replies were made. Statistics were calculated using the whole denominator (*n*), unless stated otherwise (see Fig. 1).

3 Results

3.1 Descriptives

In total, 323,512 tweets and 17,234 replies were identified and extracted for analysis (Fig. 1; Table 1). The average number of characters was 115 for the tweets (SD: 41), and 59 for the replies (SD: 38). On average, each tweet was re-tweeted 3 times (SD:

Table 1 Top-level statistics for extracted tweets and replies

	Tweet (n = 323,512)	Replies (n = 17,234)
Statistic	mean (SD)	mean (SD)
Character length	115 (41)	59 (38)
Tweet rate (per account)	10 (42)	5 (61)
Likes	3.09 (2)	0.47 (0.08)
Retweets	3 (8)	1 (2)
Hashtags	2.43 (3.07)	1.61 (0.21)
Images	0 (0.53)	0 (0.02)
Hyperlinks	0 (0.87)	0 (0.0)
Recipients (@)	1.2 (0.61)	1.24 (0.16)
Time period posted		
Morning (00:00–11:59 am)	8.34 (3.17)	2.46 (1.62)
Afternoon (12:00–11:59 pm)	16.07 (5.17)	2.01 (1.59)

SD standard deviation

8), and replies were re-tweeted on average once (SD: 2). Posters included an average of 2.43 hashtags (SD: 3.07) for tweets, and 1.61 hashtags (SD: 0.21) for replies. This study also identified that most of the tweets were posted after mid-day, i.e., between 12:00 and 11:59 pm (16.07, SD: 5.17), while most replies were sent before mid-day, i.e., between 00:00 and 11:59 am (2.46, SD: 1.62).

Using metadata for each tweet or reply, we were able to identify that the most popular methods to post a tweet or reply were Twitter for iPhone (28.55%, 21.31%), the Twitter Web Client (21.56%, 18.94%), and Twitter for Android (6.30%, 11.76%).

3.2 Sentiment Analysis

A higher proportion of tweets (48.70%; 157,576/323,512) was identified as having a positive sentiment score than a negative score (10.83%; 35,053/323,512). The reverse was found for replies, with 52.01 % identified as having a negative score (8963/17,234), and 38.78% (6683/17,234) identified as having a positive score. Quotations 1 and 2 illustrate the types of tweets and replies determined as being either positive or negative.

Figures 2 and 3 illustrate positive and neutral sentiment scores over time for tweets and replies, respectively. There was little variation in the deviation of sentiment scores over time, with no significant differences between positive and negative tweets or replies for each quarter included in the analyses. However, despite this, there was an overall downward trend in sentiment scores for tweets between Q1 2014 and Q2 2019, meaning that more negative tweets were being posted (−0.161). This was accompanied by an increase in the sentiment scores of the replies, meaning that more positive replies were being made (0.147). Although the decrease was not statistically significant, it could have been due to the higher number of tweets and replies and the rise in media attention the British Armed Forces received between 2018 and 2019, such as the investigation of former British Armed Forces personnel who served in Northern Ireland (Mills et al. 2019). Conversely, the increase in the sentiment scores of the replies during the same period could have been due to the response to negative media attention or negative Twitter content being posted. However, this increase again was not statistically significant.

Quotation 1: Example of a Positive Tweet and Reply
Tweet: "Serving in the #UK Armed Forces is the best decision I ever made and made me the man I am today #serving #military #proud"

Reply: "A big shoutout to the @ArmyLGBT team for joining @Lucianjay on air this week! It was great discussing life in the @BritishArmy and serving soldiers, who happen to be #LGBTQ. Happy Pride everyone"

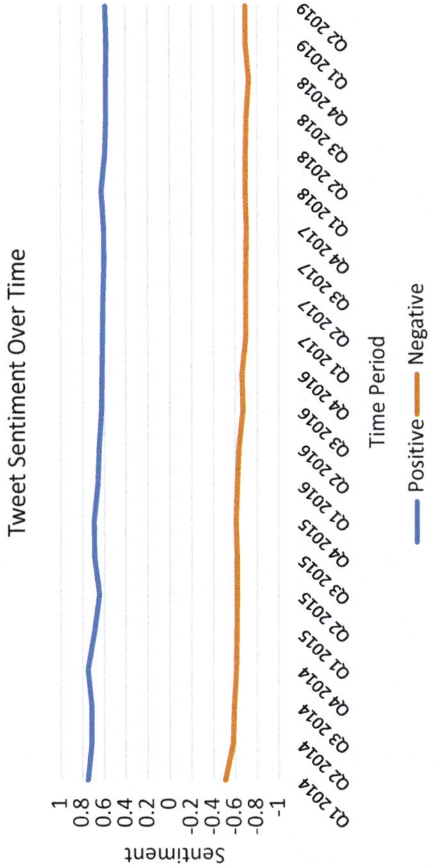

Fig. 2 Sentiment score for tweets from 1 January 2014 to 30 June 2019

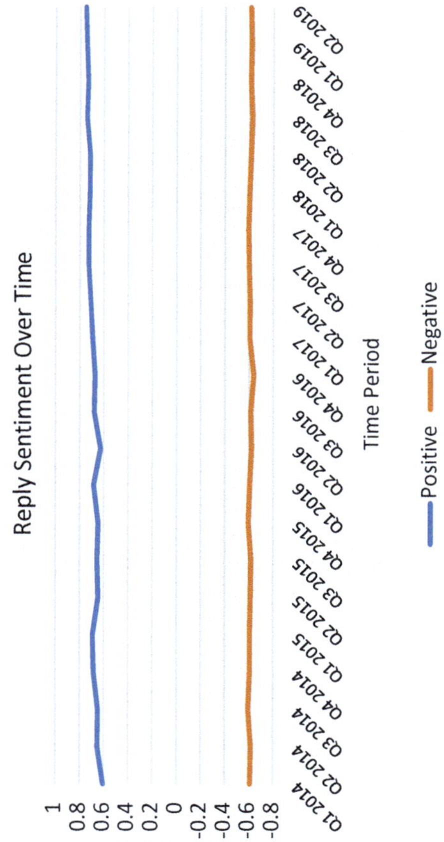

Fig. 3 Sentiment scores for replies from 1 January 2014 to 30 June 2019

Quotation 2: Example of a Negative Tweet and Reply
Tweet: "Countries would probably need less humanitarian aid if we didn't bomb them and sell weapons to countries that also bomb them #justsaying"
Reply: "@armyjobs The British government sells weapons to destabilise a foreign nation and then comes the British Army with this!"

In order to assess whether the time of day altered the sentiment score, we aggregated each tweet and reply into one minute blocks over a 24-h clock (Figs. 4 and 5). During the early hours of the day, both tweet and reply sentiment scores became more negative, whereas as the day progressed, sentiment scores improved (i.e., less negative, more positive content). Moreover, a significant difference between positive and negative tweets could be observed for the study's duration ($p = 0.045$), while significant differences were also found for tweet replies over the same time period ($p = 0.022$).

In order to further explore the role time plays for the positing of positive and negative tweets and replies, we also analysed the day of the week on which content was posted. Overall, we found no significant differences between the tweet and the reply sentiment scores (Figs. 6 and 7).

Figures 8 and 9 illustrate the sentiment score distribution of all negative and positives tweet, divided into histogram bins representing increments of *0.1* (sentiment score) to provide an overall representation of the sentiment score of all tweets and replies included in the analyses. Applying a Mann-Whitney nonparametric test revealed significant differences between all positive and negatives tweets ($p = 0.041$, $\mu = -0.04$, and SD, 0.709). This indicates that there is an overall difference between positive and negative tweets and replies. However, when stratified by time (i.e., Q1 2014), no differences exist.

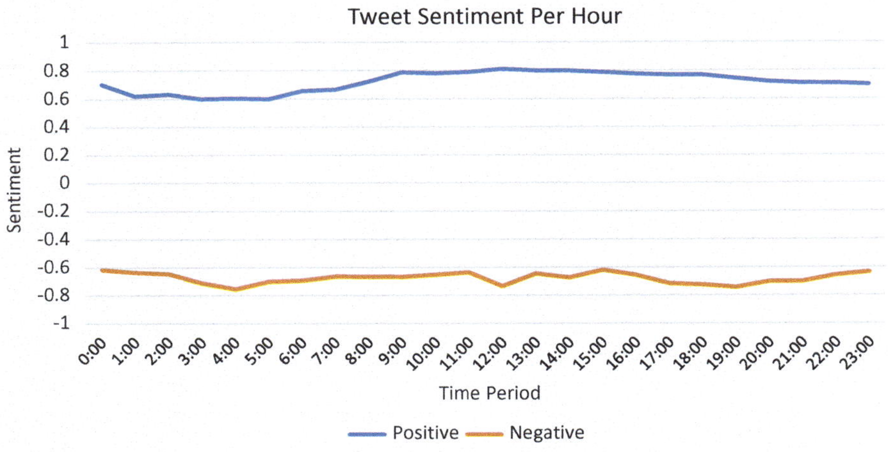

Fig. 4 Tweet sentiment score over 24 h of the day

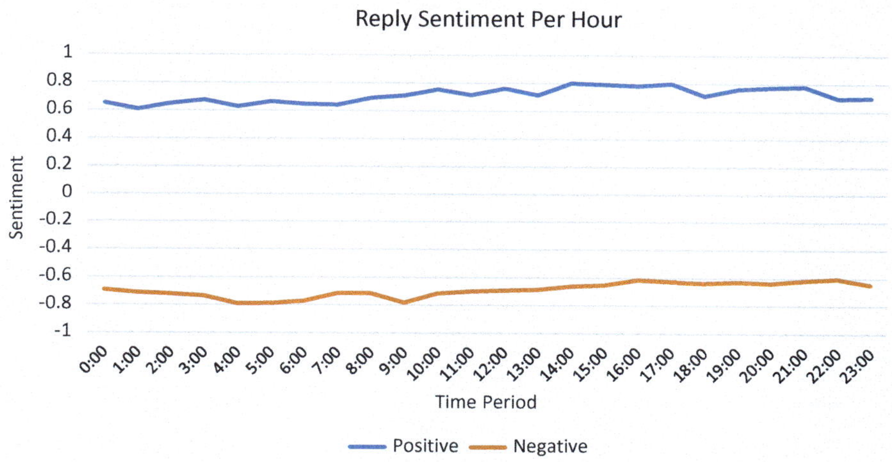

Fig. 5 Reply sentiment score over 24 h of the day

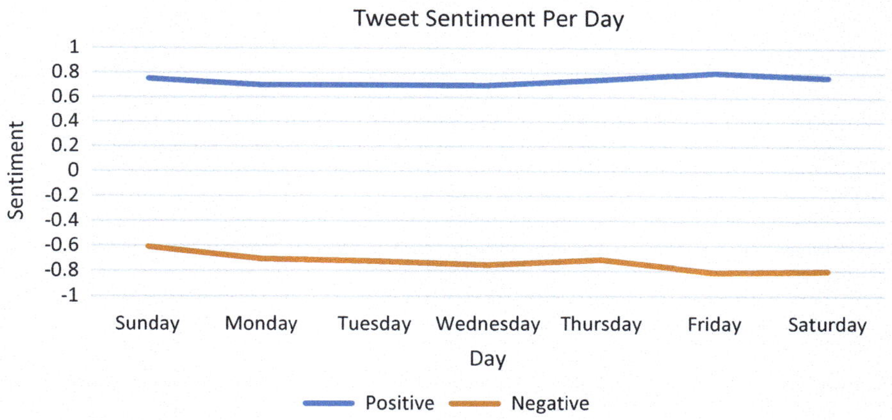

Fig. 6 Tweet sentiment score for each day of the week

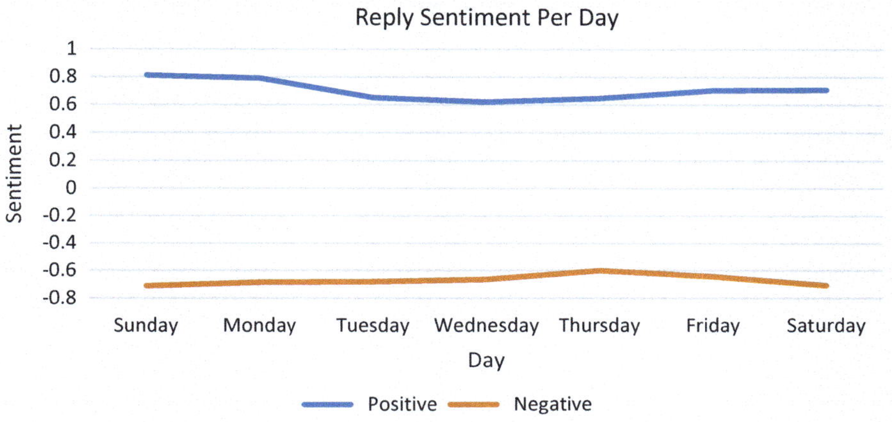

Fig. 7 Reply sentiment score for each day of the week

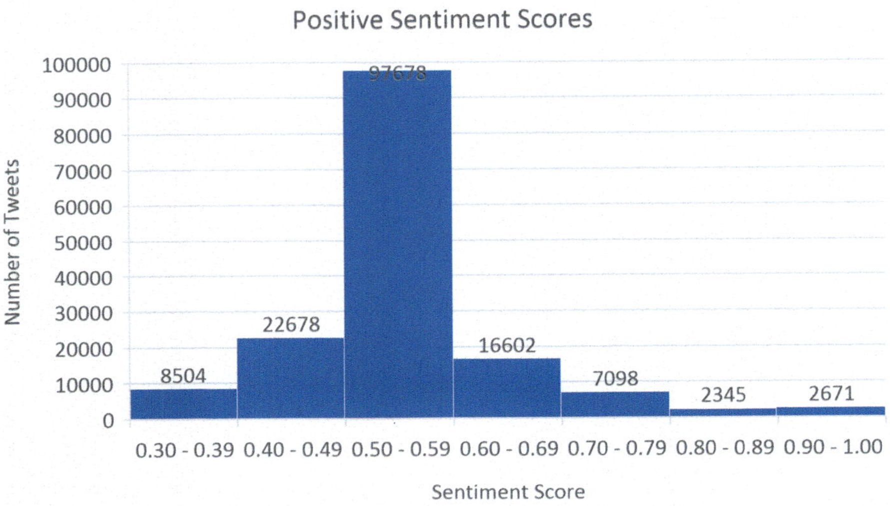

Fig. 8 Histogram spread for positive sentiment scores computed for tweets

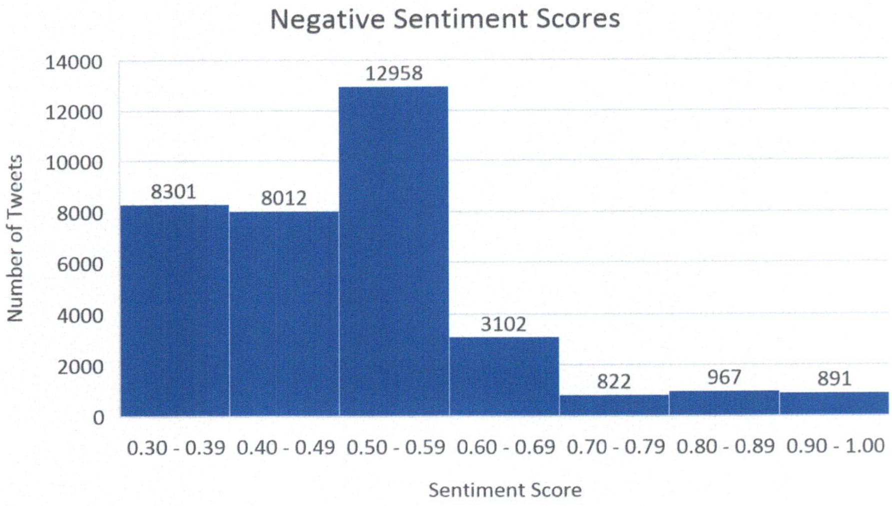

Fig. 9 Histogram spread for negative sentiment scores computed for replies

3.3 Hashtag Usage

A total of 111,014 (34.31%) tweets and 4128 replies (23.95%) included at least one hashtag (Table 2). The most popular hashtags used in the analysed tweets were '#mentalhealth' (8.27%), '#veteran' (8.14%), '#country' (3.87%), '#support' (1.24%), and '#NATO' (1.22%). The most popular hashtags in the replies were '#veteran' (12.46%), '#lestweforget' (8.68%), '#notinmyname' (5.45%), '#brexit' (4.99%), and '#labour' (2.71%). Statistically significant differences between positive and negative tweets were observed for the use of '#veteran' (44.00%, $p = 0.002$) and 'country' (49.94%, $p = 0.054$). Significant differences were also observed for tweet replies, for instance, for the hashtags '#veteran' (39.84%, $p = 0.001$), '#notinmyname' (12.97%, $p = 0.001$), and '#labour' (45.34%, $p = 0.002$).

Table 2 Top 5 hashtags computed by using the frequency of occurrence for tweets and replies

Tweet				Reply			
(n = 111,014)				(n=4128)			
Hashtag	n (%)	% positive	Chi^2 (p value)	Hashtag	n (%)	% positive	Chi^2 (p value)
#mentalhealth	9180 (8.27)	33.49	0.513 (0.473)	#veteran	514 (12.46)	39.84	14.486 (< 0.001)
#veteran	9036 (8.14)	44.00	14.466 (0.002)	#lestweforget	358 (8.68)	89.33	0.116 (0.733)
#country	4296 (3.87)	49.94	9.317 (0.054)	#notinmyname	225 (5.45)	12.97	31.158 (<0.001)
#support	1376 (1.24)	26.05	2.627 (0.267)	#brexit	205 (4.99)	48.10	9.317 (0.054)
#NATO	1354 (1.22)	58.30	2.576 (0.108)	#labour	111 (2.71)	45.34	21.278 (0.002)

3.4 Geolocation

In total, 75,766 tweets (23.41%) and 1047 replies (6.07%) had been assigned a geolocation marker, indicating latitude and longitude (Table 3). The most popular countries for posting a tweet were the UK (62.40 %), France (11.31%), Spain (7.34%), Portugal (7.19%), and Belgium (5.91%). The most popular countries for replying to a tweet were the UK (81.57%), Spain (4.10%), Belgium (3.61%), Falklands (3.44%), and USA (2.86%).

Table 3 Top 5 countries computed by using the frequency of occurrences where tweets and replies originated from according to Twitter's metadata

Tweet (n = 75,766)		Reply (n=1047)	
Country	n (%)	Country	n (%)
UK	47,277 (62.40)	UK	854 (81.57)
France	8569 (11.31)	Spain	42 (4.1)
Spain	5561 (7.34)	Belgium	37 (3.61)
Portugal	5447 (7.19)	Falklands	35 (3.44)
Belgium	4477 (5.91)	USA	29 (2.86)

4 Discussion

Overall, our findings indicate that public perceptions of the British Armed Forces were stable over time – between 2014 and 2019, no statistically significant changes were observed in the positive and negative tweets/replies analysed. These findings mirror those of the British Social Attitudes survey (Gribble et al. 2012) and illustrate that while the role of the British Armed Forces may be evolving post Iraq and Afghanistan, this appears to have little impact on perceptions, and the general public are still discussing the Armed Forces. This study did, however, find that although the day of the week a tweet or reply was posted was not significant, time of day was indeed a significant factor in determining the sentiment of a tweet or reply. Our study thus found that it was more likely for negative tweets and replies to be posted late in the evening or early morning compared to other hours of the day.

Previous research demonstrates the unpredictable and chaotic environment of social media, with a limitless variation in the content being posted at any time (Taecharungroj 2017). In this study, we found that a large proportion of tweets was identified as being positive, whereas more replies were negative. Generally, this occurred when individuals replied to a positive tweet negatively, which mimics observations made in the field of social media relations (Taecharungroj 2017). Over time this behaviour remained constant, even though the number of both positive and negative content posted increased. The unpredictable nature of Government policy and media portrayal also plays a significant role regarding the way in which the general public interact with social media. This may explain why differences were observed for the day/time of posting tweets and replies, given that news organisations are moving towards a 24-h reporting cycle and away from traditional reporting styles (i.e., reporting at 9 am).

The findings on the use of hashtags when discussing the British Armed Forces on social media provide new insight into the ways in which perceptions about the military may be politically motivated or informed through associations with current events, such as the 2016 EU referendum ('#brexit') and the 2019 UK parliamentary elections ('#labour'), and tied to particular cultural events like Remembrance Day ('#lestweforget'). This could be the result of blame, political attachment to wars, the rise of nationalism and racism due to Brexit or changes in UK Government policy regarding the armed forces (Ford and Goodwin 2017; Jennings and Stoker 2019).

Moreover, we found that the most popular hashtag associated was '#mentalhealth' for tweets, and '#veteran' for replies. We hypothesise that this may reflect the media attention on the mental health of British Armed Forces members, which resulted in an increase in public awareness regarding this issue (Ashcroft 2012; Lee 2016; Gribble et al. 2019). As UK research has shown, the public regularly overestimate the negative impacts of military service on military personnel and veterans, especially for mental health, which could explain why the majority of tweets using '#mentalhealth' were negative (Ipsos MORI 2015). Furthermore, the use of '#support' was often used in conjunction with '#mentalhealth' when tweeting. This could reinforce the notion that the public do not believe that the British Armed Forces are being supported enough as regards the mental health of their members.

A detailed analysis showed that there was a statistical difference in the use of '#mentalhealth' between positive and negative tweets, indicating a downward (more negative) focus. This could be due to users willing to be vocal on their thoughts and experiences relating to mental health and military service. Furthermore, media portrayals, such as the TV series 'Bodyguard' (Turner 2018), or veterans' stigma towards the access to and use of healthcare services could also influence public perceptions (Murphy and Busuttil 2015; Sharp et al. 2015). In this respect, it is important to acknowledge that the use of social media can give nameless individuals a vocal platform to promote their ideas and thoughts freely without balance, which could overshow and devalue other perceptions. In our analysis, we found that the most popular hashtags used were not impacted by a few 'vocal' accounts, but were dispersed, with the average number of the tweets and replies included in the study being around 9 per account.

Previous studies suggest that generally between one and two percent of tweets may contain geolocation metadata (Burton et al. 2012). However, in this study we found that approximately 24% of the tweets and 6% of the replies had a geolocation, which is significantly higher than was anticipated based on the literature. This may be due to bias, inaccurate recording or the proportion of tweets being posted by users who gave Twitter permission to collect this information. We found that the most popular country to post on the topic of the British Armed Forces was the UK. Nevertheless, interesting differences were identified between tweets and replies. While tweets were often made within countries where NATO activities were taking place, replies appear to have originated mostly from countries to which British nationals emigrate.

4.1 *Strengths and Limitations*

This study represents the first of its kind to explore the applicability of sentiment to Twitter analysis. As was the case in prior research (Gribble et al. 2012), this study found that public opinion is stable over time, with similar, albeit non-significant trends of positive and negative tweets and replies posted between 2014 and

2019. This study contributes to the academic literature by using robust and in-depth data analytics techniques to improve our understanding of public perceptions of the British Armed Forces. Furthermore, it introduces a new area of research on the politicisation of armed forces within public discourse. The method used in this study has been shown to be appropriate for this research area and could be applied in future work to provide real-time feedback on changes in recruitment strategy, mental health support, and government policy or future military operations.

The results of this study need to be considered in light of the following limitations. Firstly, this study included accounts that were manually identified. While this was systematic in nature, following the broad principles used for systematic reviews (Moher et al. 2009), future work should seek to develop more robust and automated solutions for account identification. Secondly, this study restricted the analyses to tweets and replies referring to the British Armed Forces. Thus, although the results indicated a stable trend over time, it is unknown whether this only reflects the trend of Twitter discussions on the armed forces alone. In order to address this concern, future work should extend the analysis to comparisons with other countries and occupations. Thirdly, we did not seek to separate or identify differences between active and veteran service members. Future work should therefore explore the role the active service status plays for public sentiment. Furthermore, the changes should be quantified to assess the impact positive and negative political changes, such as the EU Referendum, have on public perceptions. Finally, this study captured only a small proportion of Twitter content posted since 2014. It was limited in the number of tweets that could be collected due to Twitter API, and captured only the perceptions posted on Twitter. Moreover, the analyses were not weighted to consider the number of tweets per user account. Thus, future work should focus on the development of a longitudinal cohort of social media, specifically curated to focus on content and topics related to the British Armed Forces.

5 Conclusions

This study found that tweets related to or discussing the British Armed Forces are more positive than negative. In contrast, replies to tweets were found to be more negative than positive. For both tweets and replies, the sentiment remained stable over time, with little variation in the proportionality. In addition to sentiment scores, this study analysed the frequency of hashtags, observing variations in the use of politicised hashtags relating to Government policy or political organisations. It is not known why these differences occurred. Overall, the results of the analyses conducted for this study demonstrate that further work is required to understand the context and content of tweets and replies quantitatively and qualitatively.

References

Ashcroft M (2012) The armed forces & society: the military in Britain – through the eyes of Service personnel, employers and the public. [Online]. Lord Ashcroft KCMG PC. Available from: http://lordashcroftpolls.com/wp-content/uploads/2012/05/THE-ARMED-SOCIETY.pdf

Ashcroft M (2014) The veterans' transition review. [Online]. Lord Ashcroft KCMG PC. Available from: http://www.veteranstransition.co.uk/vtrreport.pdf

BBC (2018) Soldiers help tackle fire near Saddleworth Moor. [Online]. BBC. [Viewed 29 February 2020]. Available from: https://www.bbc.co.uk/news/uk-england-manchester-44638416

Biderman AD (1960) Morris Janowitz. The professional soldier: a social and political portrait. Ann Am Acad Pol Soc Sci 332(1):162–163. https://doi.org/10.1177/000271626033200122

Burton SH, Tanner KW, Giraud-Carrier CG, West JH, Barnes MD (2012) "Right time, right place". Health communication on Twitter: value and accuracy of location information. J Med Internet Res 14(6):e156. https://doi.org/10.2196/jmir.2121

Daniulaityte R, Chen L, Lamy FR, Carlson RG, Thirunarayan K, Sheth A (2016) "When 'bad' is 'good'": identifying personal communication and sentiment in drug-related tweets. JMIR Public Health Surveill 2(2):e162. https://doi.org/10.2196/publichealth.6327

Forces.Net (2019) Floods: British soldiers prepare to continue relief efforts throughout the night as rain persists. [Online]. Forces.Net. [Viewed 29 February 2020]. Available from: https://www.forces.net/news/warning-further-rain-british-soldiers-continue-yorkshire-flood-relief

Ford R, Goodwin M (2017) A nation divided. J Democr 28(1):17–30. https://doi.org/10.1353/jod.2017.0002

Forster A (2005) Armed forces and societies: changing roles and legitimacy. In: Forster A (ed) Armed forces and society in Europe. Palgrave Macmillan, Basingstoke

Ghani NA, Hamid S, Abaker I, Hashem T, Ahmed E (2018) Social media big data analytics: a survey. Comput Hum Behav. https://doi.org/10.1016/j.chb.2018.08.039

Gilbert E, Hutto C (2014) VADER: a parsimonious rule-based model for sentiment analysis of social media text. [Online]. In: Eighth international AAAI conference on weblogs and social media. Ann Arbor, Michigan, pp 216–255. Available from: http://www.aaai.org/ocs/index.php/ICWSM/ICWSM14/paper/download/8109/8122

Gribble R, Wessley S, Klein S, Alexander D, Dandeker C, Fear N (2012) British social attitudes: the 29th report. [Online]. In: Park A et al (eds). British social attitudes: the 29th report, pp 138–155. Available from: www.bsa-29.natcen.ac.uk/media/38852/bsa29_full_report.pdf

Gribble R, Wessley S, Klein S, Alexander D, Dandeker C, Fear N (2015) British public opinion after a decade of war: attitudes to Iraq and Afghanistan. Politics 35(2):128–150. https://doi.org/10.1111/1467-9256.12073

Gribble R, Wessley S, Klein S, Alexander D, Dandeker C, Fear N (2019) Who is a "Veteran"? RUSI J 164(7):10–17. https://doi.org/10.1080/03071847.2019.1700683

Gundlapalli AV, Carter ME, Palmer M, Ginter T, Redd A, Pickard S, Shen S, South B, Divita G, Duvall S, Nguyen TM, D'Avolio LW, Samore M (2013) Using natural language processing on the free text of clinical documents to screen for evidence of homelessness among US veterans. [Online]. AMIA Annual Symposium Proceedings 2013, pp 537–546. Available from: http://www.ncbi.nlm.nih.gov/pubmed/24551356

Hines LA, Gribble R, Wessley S, Dandeker C, Fear N (2015) Are the armed forces understood and supported by the public? A view from the United Kingdom. Armed Forces Soc 41(4):688–713. https://doi.org/10.1177/0095327X14559975

Huang M, ElTayeby O, Zolnoori M, Yao L (2018) Public opinions toward diseases: infodemiological study on news media data. J Med Internet Res 20(5):e10047. https://doi.org/10.2196/10047

Huntington S (1957) The soldier and the state; the theory and politics of civil-military relations. Am Hist Rev. Belknap Press of Harvard University Press, Cambridge. https://doi.org/10.1086/ahr/63.2.368

Ipsos MORI (2015) Hearts and minds: misperceptions and the military. [Online]. London. Available from: www.ipsos.com/ipsos-mori/en-uk/hearts-and-minds-misperceptions-and-military

Jennings W, Stoker G (2019) The divergent dynamics of cities and towns: geographical polarisation and Brexit. Polit Q 90(S2):155–166. https://doi.org/10.1111/1467-923X.12612

Jinwei C, Roussinov D, Robles-Flores JA, Nunamaker JF (n.d.) Automated question answering from lecture videos: NLP vs. pattern matching. In: Proceedings of the 38th annual hawaii international conference on system sciences. IEEE, p 43b. https://doi.org/10.1109/HICSS.2005.113

Kolliakou A, Ball M, Derczynski L, Chandran D, Gkotsis G, Deluca P, Jackson R, Shetty H, Stewart R (2016) Novel psychoactive substances: an investigation of temporal trends in social media and electronic health records. Eur Psychiatry Elsevier Masson 38:15–21. https://doi.org/10.1016/j.eurpsy.2016.05.006

Latter J, Powell T, Ward N (2018) Public perceptions of veterans and the armed forces. [Online]. London. Available from: https://www.fim-trust.org/wp-content/uploads/2018/11/20181002-YouGov-perceptions-final.pdf

Lee D (2016) Dropbox hack 'affected 68 million users'. [Online]. BBC. Available from: https://www.bbc.co.uk/news/technology-37232635

Leightley D, Pernet D, Velupillai S, Mark KM, Opie E, Murphy D, Fear NT, Stevelink SAM (2019) The development of the military service identification tool: identifying military veterans in a clinical research database using natural language processing and machine learning (Preprint). JMIR Med Inform. https://doi.org/10.2196/15852

Mahar AL, Gribble R, Aiken AB, Dandeker C, Duffy B, Gottfried G, Wessely S, Fear NT (2017) Public opinion of the Armed Forces in Canada, U.K. and the U.S. J Mil Veteran Family Health 3(2):2–3. https://doi.org/10.3138/jmvfh.3.2.002

McCartney H (2010) The military covenant and the civil-military contract in Britain. Int Aff (R Inst Int Aff 1944–) 86(2):411–428. Available from: http://www.jstor.org/stable/40664074

McCartney H (2011) Hero, victim or villain? The public image of the British soldier and its implications for defense policy. Def Secur Anal 27(1):43–54. https://doi.org/10.1080/14751798.2011.557213

McIver DJ, Hawkins JB, Chunara R, Chatterjee AK, Bhandari A, Fitzgerald TP, Jain SH, Brownstein JS (2015) Characterizing sleep issues using Twitter. J Med Internet Res 17(6):e140. https://doi.org/10.2196/jmir.4476

Mills C, Torrance D, Dempsey N, Strickland P (2019) Investigation of former armed forces personnel who served in Northern Ireland. [Online]. UK Parliament. Available from: https://researchbriefings.parliament.uk/ResearchBriefing/Summary/CBP-8352#fullreport

Moher D, Liberati A, Tetzlaff J, Altman DG (2009) Preferred reporting items for systematic reviews and meta-analyses: the PRISMA statement. PLoS Med 6(7):e1000097. https://doi.org/10.1371/journal.pmed.1000097

Morgan MJ (2001) Army recruiting and the civil-military gap. Parameters 31(2):101–117

Murphy D, Busuttil W (2015) PTSD, stigma and barriers to help-seeking within the UK Armed Forces. J R Army Med Corps 161(4):322–326. https://doi.org/10.1136/jramc-2014-000344

Pérez-Pérez M, Pérez-Rodríguez G, Fdez-Riverola F, Lourenço A (2019) Using Twitter to understand the human bowel disease community: exploratory analysis of key topics. J Med Internet Res 21(8):e12610. https://doi.org/10.2196/12610

Radzikowski J, Stefanidis A, Jacobsen KH, Croitoru A, Crooks A, Delamater PL (2016) The measles vaccination narrative in Twitter: a quantitative analysis. JMIR Public Health Surveill 2(1):e1. https://doi.org/10.2196/publichealth.5059

Rahbek-Clemmensen J, Archer EM, Barr J, Belkin A, Guerrero M, Hall C, Swain KEO (2012) Conceptualizing the civil–military gap. Armed Forces Soc 38(4):669–678. https://doi.org/10.1177/0095327X12456509

Roland D, Spurr J, Cabrera D (2017) Preliminary evidence for the emergence of a health care online community of practice: using a netnographic framework for Twitter hashtag analytics. J Med Internet Res 19(7):e252. https://doi.org/10.2196/jmir.7072

Sewalk KC, Tuli G, Hswen Y, Brownstein JS, Hawkins JB (2018) Using Twitter to examine web-based patient experience sentiments in the United States: longitudinal study. J Med Internet Res 20(10):e10043. https://doi.org/10.2196/10043

Sharp ML, Fear NT, Rona RJ, Wessely S, Greenberg N, Jones N, Goodwin L (2015) Stigma as a barrier to seeking health care among military personnel with mental health problems. Epidemiol Rev 37(1):144–162. https://doi.org/10.1093/epirev/mxu012
Strachan H (2003) The civil-military "gap" in Britain. J Strateg Stud 26(2):43–63. https://doi.org/10.1080/01402390412331302975
Szayna TS, McCarthy KF, Sollinger JM, Demaine LJ, Marquis JP, Steele B (2007) The civil-military gap in the United States: does it exist, why, and does it matter? RAND Corporation, Santa Monica. https://www.rand.org/pubs/monographs/MG379.html. Also available in print form.
Taecharungroj V (2017) Starbucks' marketing communications strategy on Twitter. J Mark Commun 23(6):552–571. https://doi.org/10.1080/13527266.2016.1138139
Turner L (2018) Bodyguard and mental health: David Budd could help save lives of those with PTSD. [Online]. BBC. [Viewed 29 February 2020]. Available from: https://www.bbc.co.uk/news/uk-45629815
Twitter (2019) Dev.twitter [online]. Twitter. [Viewed 19 June 2019]. Available from: https://dev.twitter.com/streaming/overview

A Transparent Network – Soldiers' Digital Resistance and Economic Unrest

Shira Rivnai Bahir

Abstract The current study offers several insights into the relationship between the features of digital activism and the ability of groups with limited protesting powers, such as soldiers in mandatory military service in Israel, to protest and promote social change. Moreover, it points to a unique configuration of collective identity, which is rooted not in organised collective action, but in a rhizomatic process taking place beneath the surface. The fragmented voices of these soldiers come together in the cybernetic sphere as a quasi-transparent net to form a canonical collective voice. This unique configuration seems to bridge the two existing concepts of digital activism, one of which tends to underestimate the importance of the collective, while the other believes that, despite the action of individuals, the group remains the dominant structure.

Keywords Digital activism · Collective identity · Rhizomatic activism

1 Introduction

Establishing a cross-border media system that joins the flow of information and relies on a global communications' network is linked to the change that has taken place in the activity of social movements (Van Lear and Van Aelst 2010; Chalaby 2005). This phenomenon manifested, among other things, in the mass mobilisations taking place at the beginning of the decade around the world, including extensive demonstrations in Israel. One of the prominent discussions that arose regarding the role of the internet space and social network sites in the area of activism examines the essence of collective action. In a variety of protests, a networked action structure was identified, which was devoid of any distinct leadership (McDonald 2002;

S. Rivnai Bahir (✉)
Ben-Gurion University of the Negev, Beer-Sheva, Israel

IDF Behavioral Science Center, Tel Aviv, Israel

E. Moehlecke de Baseggio et al. (eds.), *Social Media and the Armed Forces*, Advanced Sciences and Technologies for Security Applications,
https://doi.org/10.1007/978-3-030-47511-6_10

Castells 2007, 2009; Leung 2013; Bobel 2007; Fominaya 2010), but was instead driven by the actions of individuals (Bennett and Segerberg 2011, 2012; McDonald 2015; Milan 2015). In relation to these characteristics, various conceptualisations and models have been proposed to describe the flexible affinity between the individual and the collective (McDonald 2002, 2015; Hosseini 2009; Bennett and Segerberg 2012). Nevertheless, some scholars insist that it would be premature to praise the social essence of collective action in the online space (Gerbaudo 2015; Milan 2015; della Porta Pavan 2018).

Another major controversy concerns the degree of individual agency. On the one hand, some researchers believe that the characteristics of the digital space maintain power relations (Gillespie 2014; Harlow and Guo 2014; Nahon 2016; Van Dijck 2014). On the other hand, it has been argued that the online discourse sphere gives a voice as well as access to knowledge production to a wider range of population groups (Van Dijck 2009; Papacharissi 2010, 2015; Lievrouw 2011, 2018). Accordingly, a range of studies have pointed out the role of social media not only in establishing protest, but also in defining the topics of protests and empowering marginalised populations (Papacharissi and de Fatima Oliveira 2012; Sanderson et al. 2016; Gamie 2013; Breuer et al. 2015; Gabriel 2016; Thorsen and Sreedharan 2019; De Moraes et al. 2017).

The current study yields several insights into the relationship between the features of digital activism and the ability of groups with limited protesting powers, such as Israeli conscripts, to protest and promote social change. More particularly, this contribution outlines the characteristics of the soldiers of the Israel Defense Forces (IDF) as a group of activists and the way they overcome the legal restrictions they are subjected to in relation to taking part in protests.

Contrary to traditional research in the field of social movements, which focuses on a concrete movement or protest, the current research focuses on the sphere of discourse discussing national security and economic aspects on social media platforms. This empirical focus, as well as the analysis of some 600 posts written on social network sites, yielded insights on two levels – the content level, which reveals the main focus of the discussions, and the dynamics level, which reveals the identity of the speakers and the characteristics of collective identity. It appears that both the discourse and the characteristics of collective action are immersed in the socio-cultural processes taking place in Israel over the past 10 years. This also includes changes in civil-military relations. Even more so, however, the findings reveal a reflection of socio-economic perceptions and characteristics of digital activism that relate to the mass mobilisation in Israel at the beginning of the last decade.

At the content level, the study found that the discourse focused on two major issues: firstly, military service as a promise and an unrealised economic contract, and secondly, military service as an economic burden on conscripts and their parents. The fundamental discovery is the centrality of the microeconomic perspective,[1] which focuses on society's small economic units – the individual and the family.

[1] See Varian and Repcheck 2010.

This stands in contrast to the macroeconomic perspective commonly used in traditional media in relation to the debate on the defence budget.

Furthermore, the research arena was defined based on a discourse rather than a group, meaning that the identity of the speakers is not revealed. However, the discourse analysis still contributed a significant finding. The key insight is that the main topic of the discourse, whether as a speaker or as a subject, were the conscripts, a population group with limited organisational and protest power.

Finally, although – as was mentioned above – the study does not focus on a social movement or an organised protest, the existence of a recurrent canonical narrative itself nevertheless repeatedly points to the formation of a collective voice. This finding implies the formation of conscripts as a sector and as an action group, despite the restrictions imposed on them to participate in protest actions. This process of cohesion taking place below the surface and in a diffuse manner is referred to as a rhizomatic process (Castells 2015). However, both the uniqueness of the action's structure and the relationship between the individual and the collective are rooted precisely in the fact that this collective already existed as a social structure before, albeit without the ability to articulate its voice in this way. Thus, the collective is, in fact, fragmented and now reunites into a collective in action.

The present contribution presents three theoretical background overviews: social movements and the digital sphere, social movements and economic unrest in Israel, and activism and Israeli conscripts. The last two allow us to learn about the social context in which the duty soldiers' protests have evolved, both in relation to the normalisation of protests and, more specifically, regarding economic aspects and the use of social media platforms. Next, the research design and the methodology are presented. Finally, three findings are presented, two of which are concerned with content analysis, while the third examines organisation configuration and social action characteristics.

2 Social Movements and Economic Unrest in Israel

Following the global wave of protests in the Arab and the Western world, a citizen mobilisation also took place in Israel in late 2011, which included widespread activity on social networking and traditional media sites, protest camps across the country, and large demonstration marches (Swirsky 2013). Beyond being a consumer protest in favour of reducing the cost of living, this protest reflected the younger generation's sense of dissatisfaction with the neoliberal economic system and called for far-reaching changes in the economic sphere (Hershkovitz 2017). The demand for a change of priorities came alongside the expectation that the state should take care of the welfare of the individual, including providing basic services, such as education and healthcare (Grinberg 2013; Ram and Filc 2013; Rosenhek and Shalev 2014). The protesters pointed their accusations at various institutions and organisations, and the size of the defence budget was strongly criticised (Ben Hador and Rivnai-Bahir 2013).

Although the protest was widely supported by many sectors and sought to represent the 'people' as the body opposing the distribution of resources, most of its initiators and leaders were from the middle-class (Rosenhek and Shalev 2014; Ram and Filc 2013; Shenhav 2013). In this sense, the protest served as a means towards establishing a 'middle-class' social identity reaching across the familiar political, ethnic, and cultural divides (Rosenhek and Shalev 2014). At the same time, various political and ethnic identities that have shaped Israeli politics were present at a deeper level (Rosenhek and Shalev 2014; Shenhav 2013). Surveys have shown higher participation rates among those holding left-wing political views as opposed to those with right-wing views, the secular as opposed to religious people, people with a higher education as opposed to those with no higher education, and middle- and higher-income earners as opposed to low-income earners (Herman 2012). Research conducted by Rosenhek and Shalev (2014) has shown that, in many ways, these strongly represented groups are also the social groups whose economic situation has deteriorated the most over the past 20 years. Moreover, in Shalev's (2015) view, the change in the economic system was not a transition from a socialist to a neoliberal economy, but rather a change within a neoliberal system, which was mainly reflected in the removal of the protection and privileges enjoyed by certain sectors of Israel's Jewish population – including those serving in the military. With regard to the protest, it is claimed that some population groups largely refrained from participating in protest activities, such as the Palestinian population of Israel and the ultra-Orthodox (ibid.). There is a believe that these populations were even excluded from the boundaries of the collective identity at the heart of the movement, partly due to the use of a military lexicon, including completing military service as a justification for claims-making (Rosenhek and Shalev 2014; Shenhav 2013).

Opinions vary as to the results and consequences of social mobilisation that took place towards the end of 2011 in Israel (Shevchenko and Helman 2017). At the same time, however, scholars seem to agree on one key aspect: one of the main consequences was the strengthening of both the economic and the class discourse in Israeli society (Rosenhek and Shalev 2014). In this context, the protest movement challenged the assumption that resource distribution is determined by apolitical processes by way of market forces and professional economics experts (Avigur-Eshel 2018). In other words, the protest in many ways restored the political to the economic and showed how rewards, wages, and terms of employment are political ideas based on ideology rather than on some objective theory devoid of any interests or worldview (Rosenhek and Shalev 2014).

This perceptual process directly impacts the community serving in the IDF and how the recruitment model is perceived. The conscription model still largely preserves the principle of universalism, which promises great cultural and social diversity (Lomsky Feder and Ben Ari 2007). However, the fragmentation of Israeli society into subgroups and the entry into a process of hyper-sectorisation actually fractures the relationship between the Republican contract[2] and the principle of universalism. Hence, there is a growing perception that the Republican principle, which guarantees equality of resources between genders, ethno-national groups, religious groups, and classes in return for fulfilling military duty, does not exist (Levy et al. 2007).

Moreover, even if the principle of universalism seems to exist during military service, in Israel's economic reality it is dissolving. Thus, for example, the practice of prioritising the group of conscripts that does not belong to a peripheral category is slowly disappearing. At the same time, conscripts from peripheral communities who integrate into the IDF and reach senior positions enjoy the symbolic capital of military service, but join a different social and economic reality once they are released from service. This is due to the fact that the symbolic capital gained during military service is not marketable in the civilian market. In other words, the Republican entry ticket alone does not guarantee making ends meet.

3 Activism and Conscripts in the Israeli Defense Forces

According to the Israeli security service law,[3] conscription to the military service is compulsory for the entire population between the ages of 18 and 49, both for active and reserve duty. Despite the changes that have taken place in Israeli society and in civil military relations in recent decades, the military organisation is still widely appraised and trusted by the Jewish Israeli society in Israel (Tiargan-Orr and Eran-Jona 2016; Hermann et al. 2018), and military service is still perceived as an entry ticket into Israeli society (Levy et al. 2007; Lomsky-Feder and Sasson-Levy 2018). This perception also has an implication for the social movements and activism arena. The so-perceived entry ticket has turned the compulsory military service into a source of legitimisation to mobilise for a change in the social, security, and political spheres.

However, this situation contradicts the legal stipulations of the General Staff Ordinance. As is stated in the original text, soldiers are prohibited from taking part in any demonstration that is not organised by the IDF, expressing themselves publicly – orally or in writing –, or signing any petition related to any party, political or military issues.[4] As a result of this discrepancy, in most cases the activists were civilian organisations, journalists, and soldiers' families, rather than the conscripts

[2] The Republican contract refers to an approach to citizenship, which considers the military service as defining citizenship and its boundaries and sees the IDF as a state apparatus for social inclusion and exclusion (Helman 1999; Shafir and Peled 2002; Sasson-Levy 2006; Levy et al. 2007).

[3] The Defense Service Law (Consolidated Version), 5747–1986.

[4] General Staff Ordinance (1990), Order 33.0116: Public action of military personnel, public statements and relations with journalists.

themselves.[5] The few exceptions were instances of conscience objection or local protests – such as an uprising in a military correctional facility in 1997 –, many of which did not gain social support and legitimacy (Epstein 2001). Over the years, and especially after the mass mobilisation of 2011, several requests have been made to the Military Prosecutor's Office to repeal or soften the ordinance.[6] Finally, in 2016, after an appeal for letting soldiers join the Pride Parade, the wording of the ordinance was slightly softened.[7]

Starting in 2012, the IDF faced a new challenge: protests on social media. In 2014, a video was uploaded documenting a violent confrontation between an Israeli soldier and Palestinians in Hebron. The video was accompanied by a report (which later turned out to be incorrect), according to which the Israeli soldier was sent to prison following the incident. In response, a campaign erupted on the internet that sought to show solidarity with the soldier and protest his punishment. Many soldiers uploaded pictures of themselves holding the sign "I also stand with David HaNahlawi"[8] on the campaign's Facebook page as well as on other pages. Contrary to this case, most of the digital activism initiated by soldiers later did not undermine how military force was used, but dealt with the soldiers' welfare. This includes, for example, a 2015 initiative against a change in the procedure for granting conscripts a permit to grow a beard (#freewill),[9] a protest against a change in the soldiers' vacation protocol during High Holy Days of 2016 (#thesilencingprotest),[10] and a 2015 initiative that arose in protest of soldiers' wages ("But I am a poor soldier").[11]

The IDF's response to the protest phenomenon varied. In 2014, towards the end of the solidarity campaign for David HaNahlawi, Chief of Staff Benny Gantz stated that Facebook was neither a military commanding tool nor a replacement for a discussion between the commanders and their troops.[12] in the same vein, an IDF spokesman reiterated that the "IDF commands its soldiers through the commanders.

---

[5] For example, the 'Four mothers' social movement led by soldiers' mothers demanding a withdrawal of IDF forces from Lebanon, a movement led by families of war prisoners demanding their release, and a movement demanding to investigate training accidents and malfunctions led by soldiers' families (Doron and Lebel 2004).

[6] Appeal to the IDF: Rescind the ban on the participation of soldiers in demonstrations. Gili Cohen, Haaretz, 20.2.2014. The IDF changes the order: Soldiers will be allowed to attend civilian demonstrations. Gili Cohen, Haaretz, 18.07.2016.

[7] General Order 8.0105: Public activity of military personnel, public statements, and relations with journalists, the public, and the senior state level.

[8] The First Digital Revolt in the IDF, Gili Cohen and Amos Harel, 30.04.2014, Haaretz. This protest statement as well as the subsequent hashtags and comments were translated from Hebrew into English.

[9] Soldiers Protest against the 'Beard Revolution' in the IDF. Yoav Zeitun, 18.06.2015, Ynet.

[10] New Soldiers' Protest on the Net: 'Robbing Us of Our Days of Freedom'. Mor Levy, 21.03.2016, Mako.

[11] Soldiers protest: 'I'm poor, I live off my parents'. Gilad Morg, 12.11.2015, Ynet.

[12] Chief of Staff on Giant Protest: 'Facebook is Not a Commandment Tool'. Amir Bohbot, 01.05.2014, Walla.

There is no such thing as a protest on Facebook. Protest is not military."[13] Chief Education Officer Brigadier General Avner Paz-Tsuk issued a statement, according to which the social network is "a public sphere in which the military orders and procedures apply to the soldiers' activities and behaviour in the public sphere."[14] In contrast to such statements, which remained in the realm of declaration, in other cases, disciplinary measures were taken, ranging from dismissal to confinement.[15]

In 2018, the Knesset passed the Nation-State Bill.[16] This law is perceived by its opponents as being harmful to the equality between the citizens of the State of Israel as well as being a law that does not prohibit discrimination on the grounds of nationality or ethnicity.[17] Following this legislative move, the dilemma arose once again when non-Jewish soldiers and officers uploaded posts condemning the law on social networks and even calling on those belonging to the same ethnic group to stop serving in the military. In response, the IDF suspended one of the initiators and, later that year, issued an order concerning soldiers' use of the Web. Among other things, the ordinance makes it clear that "public statements on political, state or military matters" should be avoided, and that "it is forbidden to organise a protest."[18]

It is interesting to note that, in the face of this complex process, only a few researchers have considered the characteristics of the conscripts' participation in social media. As was noted by Stern and Ben Shalom (2019), the main focus of the research literature – applied, empirical, and theoretical – is on examining the characteristics of armed forces in this sphere and their ability to achieve their strategic goals with its help.

Stern and Ben Shalom's study (2019) offers a glimpse into this sphere by means of their analysis of Israeli conscripts' activity on Facebook and how this activity was perceived by the soldiers. Similarly, Kuntsman and Stein (2015), who examined a wide range of images uploaded by Israeli conscripts, found that this sphere allowed them to disseminate controversial messages anonymously without risking sanctions. At the same time, however, both studies paid little attention to the social structure of the activity or to the collective identity that emerges in this sphere.

[13] IDF Spokesman: 'There is no such thing as protests on Facebook in the IDF'. Li-Or Averbach, 01.05.2014, Globes.

[14] Chief of Staff on Giant Protest: 'Facebook is Not a Commandment Tool'. Amir Bohbot, 01.05.2014, Walla.

[15] For example: The revenge on soldiers who got stuck in the fiery heat with no air conditioning: removal and confinement for a month. Yoav Zeitun, August 12, 2015, Ynet. The soldiers who protested on Facebook were ousted from the IDF. Rafi Jarby, 15.05.2015, Maariv.

[16] Basic Law: Israel – The Nation-State of the Jewish People.

[17] Proposed Basic Law: Israel – The Nation State of the Jewish People. Adalah, 16.07.2018.

[18] General Staff Order 107.08:- Soldiers' use of the Web.

4 Social Movements and the Digital Sphere

Despite the multiplicity of theories on and definitions of social movements, there is a shared view that through organised and consistent efforts, social movements seek to bring about change within the social order (Tilly and Tarrow 2015). The mass mobilisation that took place at the beginning of the last decade in the Arab world, Europe, the USA, South America, and Israel point to the internet sphere as being a major player in this field. Without denying the understanding that these protest organisations developed in connection with local social, cultural, economic, and political transformations (Bennett and Segerberg 2012; Castells 2015; Grinberg 2013; Rosenhek and Shalev 2014; Anderson 2011; Breuer et al. 2015), many scholars have pointed out the widespread and unique use of cyber space and social networking sites as a common characteristic (Castells 2015; Bastos and Mercea 2016; Mercea 2016; Carty 2015).

Over the past decade, numerous studies have been conducted in an attempt to identify and characterise the role of the internet and social network sites in this arena, many focused on the social structure, the nature of collective identity, and the nature of the latter's formation process. Two models of collective identity have been suggested – the first one being 'platform-supported', as traditional collective action using social media sphere as a medium for organising and communicating; and the second one being 'platform-based', as collective action that takes place online and exists only thanks to the social media sphere (Karatzogianni 2018; Priante et al. 2018).

Following this, it was argued that this new organisational structure, at the centre of which stands a networked organisation, is not hierarchical, often lacks clear leadership, and shows high levels of autonomy among its members (McDonald 2002; Castells 2007, 2009; Leung 2013; Bobel 2007; Fominaya 2010). However, rather than seeing it as a collective action based on network formations, some research considers the internet organisation to be a network formation based on the activity of individuals (Bennett and Segerberg 2012; Milan 2015). In this respect, the meaning of the personalisation process is that citizens not only generate significant content, but also take part in designing and constructing the discourse and activity spheres (Papacharissi 2015). According to this model of organisation, which Milan (2013) refers to as the "cloud protesting", the cloud is a technological platform in which the cultural and symbolic activity of the group or movement is carried out with the contribution of a large number of individuals operating from their personal accounts and profiles (Milan 2013, 6).

This complexity has given rise to the question as to how collective identity can be used as the basis for action under this kind of activity and organisational structure. In which way do the dominant components function within the process of building a collective identity, including solidarity, shared values, and a sense of belonging (Leung 2013; Polletta and Jasper 2001)? Many scholars have emphasised that, in this space, the process of building a collective identity is in constant negotiation with the self in a way that constructs a flexible, fluid configuration of collective identity (Burnap and Williams 2015; Milan 2015; Bennett and Segerberg 2012; Fominaya 2010; Hosseini 2009).

Thus, the loose organisational structure is, in fact, created by means of a selective and reflexive process of representing and identifying characteristics by the individuals, the unique configuration of the structure subsequently preserving and enabling the same reflexive process (Hosseini 2009). Moreover, according to Fominaya (2010), space allows for a sense of cohesion and solidarity that develops through an open and ongoing discussion in which one can share with one another reflexive stories about a mutual activity or struggle. In order to reflect these unique characteristics, concepts such as *fluidarity* (McDonald 2002) and *interactive solidarity* (Hosseini 2009) have been proposed as substitutes for the term *solidarity* to emphasise the flexible nature of the process.

Beyond the social structure of the collective, concerns arose with regard to the nature of an individual's agency. On the one hand, there is a growing view that the Web is a technological mechanism, which develops in a space saturated with interests and actors, with most processes remaining transparent and covert (Gillespie 2014; Nahon 2016; Van Dijck 2014). According to this view, it is this very mechanism that structures collective action. There are also claims that new media is not accessible to all and that disadvantaged populations still depend on traditional means in order to convey their messages (Harlow and Guo 2014). On the other hand, it is argued that the shift from a one-way model for knowledge distribution to a multidirectional model has enabled individuals to 'talk back' in a language that is shaping cultural outcomes (Papacharissi 2010; Van Dijck 2009; Lievrouw 2011). This accessibility, without any need for specialisation, blurs the source of traditional authority, such as governments, newspaper systems, and organised commercial actors, thus contributing to its decentralisation (Papacharissi 2010; Van Dijck 2009; Lievrouw 2011; Tong and Zuo 2014). From this perspective, various researchers have shown how the sphere gives voice to marginal groups or populations that traditionally do not have access to knowledge dissemination tools. This includes, for example, women's ability to challenge social structure (Gabriel 2016; Thorsen and Sreedharan 2019; de Moraes et al. 2017), protestors' ability to mobilise in the Arab Spring (Papacharissi and de Fatima Oliveira 2012; Gamie 2013; Breuer et al. 2015), and the ability of ethnic groups to make themselves heard (Gabriel 2016; Sanderson et al. 2016). There are only a few scholars who try to crack this dichotomy. Chadwick (2017), for instance, claims that there are two systems – the former elite and the new actors –, which continue to run simultaneously. It is thus apparent that those who manage to maintain the practices of both of systems are the most empowered.

The present study hopes to yield several insights regarding both the social structure of the collective in the digital sphere and the ability of groups with limited protesting powers to protest and promote social change through said sphere.

5 Methodology

The research design was aimed at learning about the characteristics of national-security and economic discourse in the online sphere. However, the findings analysis enabled the identification of insights beyond these content aspects, as will be illustrated below.

5.1 Data Collection

This study applied a unique research design, which was based on two phases. In a first step, the economic and national security discourse arena was identified, and a discourse analysis was performed. In order to identify the boundaries of the dominant discourse arena, Facebook pages and forums that defined themselves as stakeholders in economic issues were selected based on searches on Google's generic search engine as well as on a datamining program.[19] An operator was developed, comprising 20 key words on economics (or), which were cross-referenced (and) with 20 key words on defence-related issues (or):

The key words in economics were: cost of living, budget, pension, expensive, price, the middle class, amounts, tycoons, NIS, capital, riches, money, expenses, poor, rich, cost, gaps (additional phrases with the same words). The key words on defence-related issues were: IDF, national security, soldier, army, soldiers, reserve force, weapon, war, military, uniform, security threat, generals officers, tank, missiles (additional phrases with the same words).

In the second step, among the search results, based on the volume of discourse and their activity levels, I selected the following five Facebook pages as well as three forums. The selected Facebook pages were Yair Lapid, Hagorem Ha'enoshi, Yisrael Yekara Lanu, Tzedek Hevrati – Operations Room, and Shelly Yehomovitz. The selected forums were Tapuz current events, FXP talking, and Rotter politics and current events.

The search was limited to the selected arenas over a period of 1 year (January 2014–January 2015). In addition, all chosen sites and forums had an active discourse arena[20] involving many participants.[21] A data mining software program was used to collect and monitor data.[22] Out of all the data collected on the basis of a

[19] Data monitoring used Buzilla software. http://www.buzilla.com

[20] An arena in which several posts and notifications are made daily.

[21] More than 50 participants write in addition to the forum or page administrators.

[22] Information was monitored in Buzilla, using a query that included reference to the search terms as well as the monitored sources, websites, dates, etc.

focused query on the discourse arena (approximately 9000 items), a total of 600 items)Facebook posts and blog comments(were selected for relevance.[23]

5.2 *Data Analysis*

The text corpus was analysed qualitatively to examine latent meanings and discourses. Based on a grounded theory approach developed by Glaser and Strauss (1967; Strauss and Corbin 1990; Glaser 1992), the discourse was analysed without imposing any analytical frame or hypotheses in advance. In this framework, the emphasis was put on locating different belief and language patterns (Johnstone 2018, 3) in relation to the link between national security, military service, and economic aspects.[24]

6 Findings

The analysis of the research findings provides insights into the collective action of soldiers on two levels: the content of the discussions and the relationship between the individual and the collective, including the identity of speakers.

6.1 *The Unfulfilled Republican Contract*

The discourse on the Net points to a focus on material retribution as the basis for ensuring socio-economic resilience, rather than focusing directly on retribution as a contribution to State security. This perspective does not refer solely to the immediate material remuneration of the conscripts, but expands and extends to their economic rights after they complete their military service.

Several people have uploaded posts criticising the cost of living in Israel, including real estate prices. These arguments were raised in relation to the link between fulfilling military service duty and the inability to achieve economic well-being in Israel. Thus, the socio-economic position, as expressed in relation to the remuneration of conscripts, becomes, in fact, completely subordinated to the Republican principle – i.e., the State must look after the needs and welfare of those who contributed to the State by serving in the military. The following post was written by a soldier's parent:

---

[23] The relevance index is an index that orders the search results by their relevance to the query, based on topic salience, the existence of many the search terms, frequency of search terms, etc. This is a built-in index of Buzilla.

[24] All data was translated into English.

> I, who educated my children to be loyal to our country, fighters in the army, I'm not surprised that the best youth feel alienated and don't want to serve. The children can't see the day-to-day coping and hear the words 'there's nothing' ... Please, this is a cry for help for change, for living in dignity!

A similar statement was posted by a former soldier:

> You get up one month after your release from the army and you have no direction at all in life, and now you know that you need to save money, because to take a mortgage in Israel you need 30% equity !!!

A leading principle in the security ethos is the need to ensure the continued existence of the State of Israel, where security existence refers, inter alia, to the possibility of granting citizens a certain quality of life. One of the most poignant arguments in this regard concerns the fact that there is no connection between watching over the land in the security sense and the land as a place of sustenance, as a resource, and an asset: "I fought on this land, and when I ask to live in it, unfortunately it's not possible."

Military service, therefore, is presented in the discourse as the basis for legitimising claims concerning economic rights. The allegations of unfulfilled promises largely reflect a crack in or disappointment with the Republican contract. The basic premise of these claims is that the relationship between military service and belonging to society carries with it a promise of – albeit limited – economic well-being. The moderate voice of this discourse calls upon the State's various institutions to care for its citizens by providing services, such as healthcare and education, and by managing an economic system that will ensure their wellbeing. Hence, the main demand, similar to the 2011 protest, is to lower the cost of living and housing prices. The more radical discourse even attributes malice to the breach of contract, actually making allegations of deliberate exploitation of the conscripts. In this context, there is an explicit call for citizens to also breach the contract by refusing to enlist or refusing to serve in combat positions, and to choose positions with a lesser level of commitment and intensity. In both the moderate and the radical discourses, the economic levy on combat service is perceived as being extremely high. In other words, from an economic point of view, it is the least worthwhile transaction, as it involves high expenses, but does not allow one to work or acquire a profession at the same time as doing military service, while the experience gained does not constitute an advantage on the civilian labour market.

6.2 *Carrying the Economic Burden*

One of the main arguments that resonated in this study on the network discourse regarding the model of compulsory service was linked to soldiers' economic situation, as well as that of their families. Analysing the discourse on the Web has strongly raised these arguments, pointing to a real difficulty in the ability of conscripts to live well and pay for their basic needs. Writers on the internet claim that soldiers' wages/subsistence wages are insufficient and thus require a widespread use of alternative sources (savings, parents' support, work, gifts). These claims relate both to paying for soldiers' current and daily needs (including providing the equipment they need to fulfil their duties) and to financing 'luxuries', such as phone bills and entertainment.

The following quotes are two examples of posts referring to these issues. The first one was written by a soldier, the second by a soldier's mother:

> The equipment is expensive. We can't afford to buy it ourselves and no one cares. Please help us keep warm.

> My son was also in combat engineering, and if I hadn't provided him with everything the IDF didn't provide, my son would be in the same situation as these soldiers.

Furthermore, some of the writers claim that soldiers and their families were explicitly asked by their commanders to purchase the necessary equipment with their own money.

The fact that the soldier must be supported financially, as well as the economic reality he/she will have to face in the future, is perceived as being harmful to him/her and his/her family. More importantly, compared to other, non-mobilised population groups, the problem – and with it the claim of inequality in the distribution of the economic burden at the individual and State levels during and after military service – intensifies. The following example illustrates this point:

> How do you help a child who's joining the army … so that he won't lack anything? You have to take extra work, but then you have to pay additional tax. And we miss the financial help he would have got with his work … The child goes to serve the State, we're doing everything we can to buy him suitable clothing and to give what the army doesn't supply, while entire sectors are not serving, are working, are harming the State's security and no one has an answer. What should we do????

Apart from the lack of funding itself, commentators speak of an unfair distribution of resources due to the over-budgeting of yeshiva students,[25] to their not carrying the tax burden, and to the ultra-Orthodox population's avoidance of joining the labour market. This is elaborated in the following statement:

> I see thousands of soldiers who spend days and nights in the Territories to protect us!!! The same soldiers who, once they are released, will be unemployed, they won't have a penny in their pocket. … These dead soldiers whose families have collapsed. And on the other hand, I see all the benefits that go to ungrateful people who didn't even think of joining the army or contributing to the State in any way. I say that we're all equal in rights and obligations!!!! Whoever gives should get in return!!!

The discourse on fairness, which is at the basis of the arguments concerning the lack of equality, is dominant and central among youth and, especially, among the candidates for security service (Rivnai Bahir and Avidar 2017). It is rooted in an ongoing debate regarding the implementation of the recruitment model among yeshiva

---

[25] The term yeshiva students refers to young ultra-Orthodox Jews in Israel who study in Jewish religious educational institutions (yeshiva).

students. According to a historical arrangement, yeshiva students may be exempt from military service to allow them to devote their time to religious studies. This agreement, although contradictory to the Defense Service Law,[26] has existed since the formation of the State of Israel and therefore constitutes a legal and public controversy. This controversy is partly responsible for the lack of political stability that has characterised the State of Israel in the past decade.[27]

Given the double burden and the focus on the economic dimension, this type of non-recruitment is not necessarily the main factor for the perceived lack of equality of sharing the burden. In that regard, compulsory mobilisation for the ultra-Orthodox population is not seen as the ultimate solution because of the assumption that an ultra-Orthodox soldier will receive a much higher subsistence fee/salary. This is made clear in the following statement:

> It makes no sense to recruit an ultra-Orthodox citizen at the age of 21, since that same citizen will already be married and have a few children. Therefore, not only will that citizen not contribute significantly to any sector in which he performs his duty, but [he] will also cost the country far more than any other soldier or person serving – since the family of that ultra-Orthodox soldier must be supported. This will put a strain on the defence budget and, of course, create wage inequality, whereas the contribution of that ultra-Orthodox soldier with a number of children will be marginal. With all due respect, the IDF soldiers 'work their butts off' during their military service, and a father of several children will not be able to serve in the same way – and since his salary will be much higher than that of a regular soldier who contributes more, there will be an increase in the deficit of the defence system's budget.

The analysis of the conversations appearing on the internet therefore shows that the centre of gravity of the claim regarding a lack of fairness is not found primarily in the actual non-serving reference group, but rather in the rewards of those who are eligible. Thus, the debate on the burden of serving in the military also rests on the notion that the Republican contract between the State and the individual is unfulfilled. As one person wrote on Yair Lapid's Facebook page:

> So, when will you get rid of this 'equality of burden' nonsense … and begin to improve the conditions of the soldiers who are already serving and contributing? … There are bases where food is poor, living conditions are not good. The salaries of all the soldiers are simply shameful, the grants are low. There are not enough scholarships and assistance after the army. Help them during the military to complete and improve their maturity exams, pass the psychometric, study a profession, … any certificate.

The writer goes on to emphasize that the equality of burden, from the perspective of material rewards, is an unattainable goal (even among those who serve in the military) and should therefore not be pursued at all:

> You want to bring another 2,000–3,000 special soldiers, do everything for them to make the army suitable for them, pay them high salaries – for what? For a non-existent equality of burden? There is not, and will never be, such a thing. Even regular soldiers serve differently. One leaves at 3 pm and goes to work, the other stays at the base for 21 days. Around 45%

[26] The Defense Service Law (Consolidated Version), 5747 (1986).

[27] See Stadler 2007.

> of women in Israel do not serve in the army. Where's the equality here? There are soldiers who serve 17 months compared to soldiers who serve 36.

As can be deduced from these statements, the economic-security discourse in social media favours the principle of functional compensation over the principle of fairness and universalism. The meaning of this alternative principle is that giving high rewards to those who perform a role is vital and essential for society.

7 Transparent Collective in Action

As regards the transparent collective in action, the research design did not include a focus on the group or the collective. Nevertheless, the analysis of the discourse arena helped identify the dominance of conscripts – present or past – both as a subject and as a speaking group. Their dominant role in the discourse may be attributed to a range of factors, from the normative nature of social media use by the members of this age group, to the size and breadth of the population of IDF soldiers in the past and present, to the fact that Israel has military conscription, as a result of which the IDF is the people's army and evokes Israeli society's deep identification with soldiers and their families (Rivnai Bahir and Avidar 2017).

However, it seems that, most of all, the actual characteristics of the field – i.e., the social media platforms – enabled the creation of conscripts as generators of discourse. The 'voice' of the conscripts analysed in this research is, in fact, a collection of individual voices not associated with any organised protest action structure or unionised movement calling for change; there was no leading protest page or sweeping hashtag. This finding is consistent with the view that has emerged in the literature[28]: the collective framework in the social media space is no more than a network of individuals acting as individuals.

In addition, as is reflected in the content of the posts, the voices that gather on the Web into a single canonical voice are expressing a discomfort. The discomfort raised in social media focuses on microeconomic aspects, such as the quality of life and economic condition of the individuals – i.e., the soldiers. Contrary to this, in recent years, the discourse on economic-national security in the mass media has allocated much space to the macroeconomic perspective, namely, to discussions on the defence budget. In this context, the focus was on the size of the budget in connection with economic growth and national priorities.

This shift seems to be related to two processes. Firstly, the unique characteristics of social media – most notably the personalisation of social phenomena – also permeate the online debate on compulsory military service and soldiers, focusing on the microeconomic level as well as on everyday life. Secondly, the place of the economic dimension and discourse in the public sphere and on the agenda has expanded over the years, bringing with it a plethora of connections. It is no longer a

[28] As was detailed in the theoretical review of this contribution.

niche discourse for connoisseurs and economists, but a widespread discussion that takes place in a multitude of spheres, through a wide range of actors. This phenomenon is, in part, due to the fundamental change that has taken place in the processes of knowledge production, dissemination, and consumption over the past 10 years. Thus, the transformation of the public discourse into a personal-individual discourse marks another stage in the expansion of the economic discourse on armed forces and security and the arrival of new voices.

The emergence of such a socio-economic discourse is linked to the protests that took place in Israel and the Western world at the beginning of the past decade. As was mentioned above, the middle class, which was at the centre of the protests taking place in Israel in 2011, not only had economic class affiliation, but also incorporated within it a political and ethnic affiliation that was reflected, among other things, in the use of the Republican contract (Shenhav 2013; Rosenhek and Shalev 2014; Ram and Filc 2013). The basis of the contract, as previously discussed, is the burden of security, while the core population bearing it, – in the past as well as in the present – are the conscripts. Therefore, the renewed intensification of conscripts' instrumental claims analysed in this study can be seen as a local expression of the 2011 protest.

8 Conclusion

The study in this contribution reveals a change in the dialectic relationship between the individual and the collective. In the analysed case, soldiers' activity in the digital space enables the creation of complex relations of separation and reconnection between the individual and the collective as well as the collective identity. The conscripts form a cohesive collective with clear and distinct boundaries. At the same time, because of military orders, their ability to take part in protest activity is significantly limited, and they cannot become an acting collective. Therefore, the existing uneasiness is uploaded onto the digital space as individual voices seeking to carry out change.

This process is consistent with the understanding of the personal nature of action in the digital space and, in particular, through social media (Bennett and Segerberg 2011, 2012; Papacharissi 2015). According to Bennett and Segerberg (2011, 2012), in this situation, the logic of action does not rest on collectivity, but rather on connective action. However, the uniqueness of the present case goes beyond this notion, emphasising that the same collective actually already existed as a social structure, but without being able to make itself heard in this way. Although the initial affiliation with the collective was and remains legitimate, given the limitations, it does not form a basis for action or an anchor for organising, but rather splits into a number of individual voices. In connection with the characteristics of the digital space, these individual voices are reunited from the other side into a renewed collective identity with different and unique characteristics. Castells (2015, 15), who followed the wave of protests in 2011, claims that the rallying of voices in the social media space is a rhizomatic social movement. Drawing on biological conceptualisation, which describes how roots propagate,[29] Castells describes the process as a horizontal rather than a hierarchical process.[30]

Indeed, the way in which the participants in this discourse act and emerge, as is reflected in the analysis of the findings of this study, shows that the discourse or argument is not hierarchically structured, but emerges horizontally in a number of places. Thus, the collective actually splits and breaks up into a variety of voices on different channels, after which, through a rhizomatic process, it once again becomes a collective in action.

References

Anderson L (2011) Demystifying the Arab Spring: parsing the differences between Tunisia, Egypt and Libya. Foreign Aff 90(3):2–7

Avigur-Eshel A (2018) More of the same: discursive reactions of members of Knesset to the 2011 social protest in Israel. Middle East Law Gov 10(2):117–140

Bastos MT, Mercea D (2016) Serial activists: political twitter beyond influentials and the twittertariat. Mew Media Soc 18(10):2359–2378

Ben Hador B, Rivnai-Bahir S (2013) "The defense budget is a fog screen, so we will not see what happens there" – the discussion on the network about the defense budget during the 'tent protest'. Lecture at the 44th conference of the Israeli Sociological Association. Ruppin Academic Center [Hebrew]

Bennett LW, Segerberg A (2011) Digital media and the personalization of collective action. Inf Commun Soc 14(6):770–799

Bennett LW, Segerberg A (2012) The logic of connective action. Inf Commun Soc 15(5):739–768

Bobel C (2007) 'I'm not an activist, though I've done a lot of it': doing activism, being activist and the 'perfect standard' in a contemporary movement. Soc Mov Stud 6(2):147–159

Breuer A, Landman T, Farquhar D (2015) Social media and protest mobilization: evidence from the Tunisian revolution. Democratization 22(4):764–792

Burnap P, Williams ML (2015) Cyber hate speech on twitter: an application of machine classification and statistical modeling for policy and decision making. Policy Internet 7(2):223–242

Carty V (2015) Social movements and new technology. Westview Press, Boulder

Castells M (2007) Communication, power and counter-power in the network society. Int J Commun 1(1):238–266

Castells M (2009) Communication power. Oxford University Press, Oxford

Castells M (2015) Networks of outrage and hope: social movements in the internet age, 2nd edn. Polity, Cambridge

Chadwick A (2017) The hybrid media system: politics and power. Oxford University Press, Oxford

Chalaby JK (2005) From internationalization to transnationalization. Glob Media Commun 1(1):28–33

---

[29] For more information, see Cheol Seong Jang et al. (2006).

[30] Although Castells (2015) mentioned this concept in relation to social movements, a few decades earlier, Deleuze and Guattari (1987 [1980]) used this metaphor to describe their theoretical ideas on the formation of political thought. In this theory, the emphasis was on the absence of a central source in or a central direction to the plethora of distributed ideas. Similarly, Latour (1999, 19) used this concept to describe the characteristics of the Web as loose and changing. See also Jensen (2019).

De Moraes GHSM, Boldrin J, Silva DS (2017) Participation in Brazilian feminist movements on social networks: a study on the campaign Meu Amigo Secreto (my secret Santa). Inf Soc 27(2):219–234. [Portuguese]

Deleuze G, Guattari F (1987) A thousand plateaus. University of Minnesota Press, Minneapolis

Della Porta D, Pavan E (2018) The nexus between media, communication and social movements looking back and the way forward. In: Meikle G (ed) The Routledge companion to media and activism. Routledge, New York, pp 29–37

Doron G, Lebel U (2004) Penetrating the shields of institutional immunity: the political dynamic of bereavement in Israel. Mediterranean Politics 9(2):201–220

Epstein AD (2001) Mutiny in the 6th military jail and the patterns of public protest in a changing Israeli society. Democr Cult 4–5:13–54

Fominaya CF (2010) Creating cohesion from diversity: the challenge of collective identity formation in the global justice movement. Sociol Inq 80(3):377–404

Gabriel D (2016) Blogging while black, British and female: a critical study on discursive activism. Inf Commun Soc 19(1):1622–1635

Gamie S (2013) The cyber-propelled Egyptian revolution and the de/construction of ethos. In: Folk M, Apostel S (eds) Online credibility and digital ethos: evaluating computer-mediated communication. IGI Global, Hershey, pp 316–330

Gerbaudo P (2015) Protest avatars as memetic signifiers: political profile pictures and the construction of collective identity on social media in the 2011 protest wave. Inf Commun Soc 18(8):916–929

Gillespie T (2014) The relevance of algorithms. In: Gillespie T, Boczkowski PJ, Foot A (eds) Media technologies: essays on communication, materiality and society. MIT Press, Cambridge, MA, pp 167–194

Glaser B, Strauss AL (1967) The discovery of grounded theory, Chicago: Aldine

Glaser BG (1992) Basics of grounded theory analysis: Emergence vs. forcing. Mill Valley. CA: Sociology Press

Grinberg LL (2013) The j14 resistance mo(ve)ment: the Israeli mix of Tahrir Square and Puerta del Sol. Curr Sociol 61(4):491–509

Harlow S, Guo L (2014) Will the revolution be tweeted or facebooked? Using digital communication tools in immigrant activism. J Comput-Mediat Commun 19(3):463–478

Helman S (1999) Negotiating obligations, creating rights: conscientious objection and the redefinition of citizenship in Israel. Citizsh Stud 3(1):45–70

Hermann T (2012) The Israeli democracy index 2012. The Israeli Democracy Institute, Jerusalem. [Hebrew]

Hermann T, Anabi O, Heller E, Omar F (2018) The Israeli democracy index 2018. The Israel Democracy Institute, Jerusalem. [Hebrew]

Hershkovitz S (2017) "Not buying cottage cheese": motivations for consumer protest – the case of the 2011 protest in Israel. J Consum Policy 40(4):473–484

Hosseini SAH (2009) Alternative globalizations: an integrative approach to studying dissident knowledge in the global justice movement. Routledge, New York

Jang CS, Kamps TL, Skinner DN, Schulze SR, Vencill WK, Paterson H (2006) Functional classification, genomic organization, putatively cis-acting regulatory elements, and relationship to quantitative trait loci, of sorghum genes with rhizome-enriched expression. Plant Physiol 142(3):1148–1159

Jensen CB (2019) Is actant-rhizome ontology a more appropriate term for ANT? In: Blok A, Farias I, Roberts C (eds) The Routledge companion to actor-network theory. Routledge, New York, pp 73–86

Johnstone B (2018) Discourse analysis, vol 3. Wiley, Hoboken

Karatzogianni A (2018) Leaktivism and its discontents. In: Meikle G (ed) The Routledge companion to media and activism. Routledge, London, pp 250–258

Kuntsman A, Stein RL (2015) Digital militarism: Israel's occupation in the social media age. Stanford University Press, Stanford

Latour B (1999) On recalling ANT. In: Law J, Hassard J (eds) Actor-network theory and after. Blackwell Publishers, Oxford, pp 15–25
Leung A (2013) Anonymity as identity: exploring collective identity in anonymous cyberactivism. Int J Technol Knowl Soc 9(2):173–184
Levy Y, Lomsky-Feder E, Harel N (2007) From "obligatory militarism" to "contractual militarism" – the changing face of militarism in Israel. Isr Stud 12(1):127–148
Lievrouw L (2011) Alternative and activist new media. Polity Press, Cambridge
Lievrouw LA (2018) Alternative computing. In: Meikle G (ed) The Routledge companion to media and activism. Routledge, New York, pp 65–73
Lomsky-Feder E, Ben-Ari E (2007) Diversity in the Israel defense forces. In: Soeters JL, Van der Meulen J (eds) Cultural diversity in the armed forces. Routledge, Abingdon and New York, pp 139–153
Lomsky-Feder E, Sasson-Levy O (2018) Women soldiers and citizenship in Israel: gendered encounters with the state. Routledge, Abingdon
McDonald K (2002) From solidarity to fluidarity: social movements beyond "collective identity"– the case of globalization conflicts. Soc Mov Stud 1(2):109–128
McDonald K (2015) From Indymedia to anonymous: rethinking action and identity in digital cultures. Inf Commun Soc 18(8):968–982
Mercea D (2016) Civic participation in contentious politics. Palgrave, Basingstoke
Milan S (2013) WikiLeaks, anonymous, and the exercise of individuality: protesting in the cloud. In: Brevini B, Hintz A, McCurdy P (eds) Beyond WikiLeaks: implications for the future of communications journalism and society. Palgrave Macmillan, Basingstoke, pp 191–208
Milan S (2015) When algorithms shape collective action: social media and the dynamics of cloud protesting. Soc Media Soc 1(2):1–10
Nahon K (2016) Where there is social media there is politics. In: Bruns A, Enli G, Skogerbo E, Larsson AO, Christensen C (eds) The Routledge companion to social media and politics. Routledge, New York, pp 39–55
Papacharissi Z (2010) A networked self. In: Papacharissi Z (ed) A networked self: identity, community, and culture on social network sites. Routledge, New York, pp 304–318
Papacharissi Z (2015) Affective publics: sentiment, technology, and politics. Oxford University Press, New York
Papacharissi Z, de Fatima Oliveira M (2012) Affective news and networked publics: the rhythms of news storytelling on# Egypt. J Commun 62(2):266–282
Polletta F, Jasper JM (2001) Collective identity and social movements. Annu Rev Sociol 27(1):283–305
Priante A, Ehrenhard ML, van den Broek T, Need A (2018) Identity and collective action via computer-mediated communication: a review and agenda for future research. New Media Soc 20(7):2647–2669
Ram U, Filc D (2013) The 14th of July of Daphni Leef: the rise and fall of the social protest. Theory Criticism 41:17–43. [Hebrew]
Rivnai Bahir S, Avidar M (2017) Alternative vs. canonical discourses regarding military service. In: Tibor ST, Eva MB (eds) Recruitment & retention, part 2. Res Militaris, 5 [online]. Available from: http://resmilitaris.net/index.php?ID=1025992
Rosenhek Z, Shalev M (2014) The political economy of Israel's 'social justice' protests: a class and generational analysis. Contemp Soc Sci 9(1):31–48
Sanderson J, Frederick E, Stocz M (2016) When athlete activism clashes with group values: social identity threat management via social media. Mass Commun Soc 19(3):301–322
Sasson-Levy O (2006) Identities in uniform: masculinities and femininities in the Israeli military. Magnes, Jerusalem. [Hebrew]
Shafir G, Peled Y (2002) Being Israeli: the dynamics of multiple citizenship, vol 16. Cambridge University Press, Cambridge

Shalev M (2015) From socialism to capitalism? Economy and society in Israel. Lecture at a conference in the study of Israeli society to mark the 65th anniversary of the journal "Megamot". [Hebrew]
Shenhav Y (2013) The carnival: protest in a society without oppositions. Theory Criticism 41:121–145. [Hebrew]
Shevchenko Y, Helman S (2017) Anti-neoliberal protest and neoliberal outcomes: the appropriation and translation of the 'Tents Protest' by the Trachtenberg Committee. Isr Sociol 19(1):145–168
Stadler N (2007) Playing with sacred/corporeal identities: Yeshivah students' fantasies of military participation. Jew Soc Stud 13(2):155–178
Ster N, Shalom UB (2019) Confessions and tweets: social media and everyday experience in the Israel Defense Forces. Armed Forces Soc:1–24
Strauss AL, Corbin JM (1990) Basics of qualitative research: grounded theory procedures and techniques. Sage, Sage Newbury Park
Swirsky S (2013) The people demand control over their money. Theory Criticism 41:147–163. [Hebrew]
Thorsen E, Sreedharan C (2019) #EndMaleGuardianship: women's rights, social media and the Arab public sphere. New Media Soc 21(5):1121–1140
Tiargan-Orr R, Eran-Jona M (2016) The Israeli public's perception of the IDF: stability and change. Armed Forces Soc 42(2):324–343
Tilly C, Tarrow S (2015) Contentious politics, 2nd edn. Oxford University Press, New York
Tong J, Zuo L (2014) Weibo communication and government legitimacy in China. Inf Commun Soc 17:66–85
Van Dijck J (2009) Users like you? Theorizing agency in user-generated content. Media Cult Soc 31(1):41–58
Van Dijck J (2014) Datafication, dataism and dataveillance: big data between scientific paradigm and ideology. Surveill Soc 12(2):197
Van Lear J, Van Aelst P (2010) Internet and social movement action repertoires: opportunities and limitations. Inf Commun Soc 13(8):1146–1171
Varian HR, Repcheck J (2010) Intermediate microeconomics: a modern approach, vol 6. WW Norton & Company, New York

Part IV
Risks and Dangers of Social Media

The Dark Side of Interconnectivity: Social Media as a Cyber-Weapon?

Sofia Martins Geraldes

Abstract Traditional wisdom understands weapons as tools that cause or have the potential to cause damage or harm, whereas cyber-weapons relate to the use of computer code that causes or has the potential to cause damage or harm. Both conceptions understand damage and harm as inherently physical. While the rise of social networks creates new opportunities for strategic communications in the armed forces, it also facilitates hostile activities, such as psychological operations, with the potential to cause damage beyond the physical domain, thus challenging the traditional understanding of weapons. This contribution investigates the potential of social media to be used as a cyber-weapon, arguing that Russia used social media as a cyber-weapon in the conflict with Ukraine. The analysis demonstrates that Russia's use of social media caused damage to Ukraine, which consequently contributed to the reform of the security and defence sector in Ukraine.

Keywords Cyber-weapons · Social media · Psychological operations · Armed forces · Russia-Ukraine conflict

1 Introduction

Despite the momentum that emerging cyberspace-related threats and challenges have been gaining on the post-Cold War international security agenda as a result of technological innovations and changes in the geopolitical landscape, cyberspace was not always considered a security issue (Maness and Valeriano 2015; Hansen

The author would like to thank the reviewers for their excellent comments.

The present research is funded by Fundação para a Ciência e a Tecnologia (SFRH/BD/140797/2018).

S. M. Geraldes (✉)
Instituto Universitário de Lisboa (ISCTE-IUL), Centro de Estudos Internacionais, Lisbon, Portugal
e-mail: Sofia_Cristina_Geraldes@iscte-iul.pt

E. Moehlecke de Baseggio et al. (eds.), *Social Media and the Armed Forces*, Advanced Sciences and Technologies for Security Applications,
https://doi.org/10.1007/978-3-030-47511-6_11

and Nissenbaum 2009; Eriksson and Giacomello 2006). This situation is reflected on the incipient conceptualisation of what counts as a cyber-threat, a cyber-attack, and especially a cyber-weapon.

The term cyber-weapon is yet to be given a consensual definition and, lacking analytical rigour, has thus been used as a catch-all term for different forms of hostile use of cyberspace (Stevens 2017). In this context, Rid and McBurney (2012, 7) attempted to define what does or not count as a cyber-weapon, understanding it as "a computer code that is used, or designed to be used, with the aim of threatening or causing physical, functional, or mental harm to structures, systems or living beings".

This conceptualisation contributes to the narrow understanding of the forms the hostile use of cyberspace can take, which is generally associated with the disruption of critical infrastructures. However, the Russia-Ukraine conflict demonstrated that the hostile use of cyberspace can also take place on social media through different types of military operations, where the main targets are not machines or networks but minds (Lange-Ionatamishvili and Svetoka 2015). While a cyber-attack against critical infrastructures can have devastating effects, an attack on cyberspace can also present itself through the exploitation of the digital domain to shape perceptions, influence decision-making, and provoke actions on its target (Lange-Ionatamishvili and Svetoka 2015). According to Lange-Ionatamishvili and Svetoka (2015), the consequences of this type of cyber-attack can be as severe as those of an attack on critical infrastructures. This represents what Herrick (2016, 99) termed "the social side of 'cyber power'".

Furthermore, the debate on social cyber-attacks and the hostile use of social media has been centred on its use in peace-time situations, focussing mostly on its effectiveness as a tool to disseminate disinformation that seeks to manipulate the population of the adversary, and as a challenge to democracy.[1] However, social media has also been used in conflict situations and against the armed forces with the aim of damaging the moral of the adversary, which in turn can damage the fighting capacity and therefore the capability to respond to aggression (Mölder and Sazonov 2018).

With regard to this scenario, the aim of this paper is to analyse the potential of social media to be used as a cyber-weapon in interstate conflict situations, namely by and against the armed forces, by means of a case-study analysis of the Russia-Ukraine conflict.[2] The central argument is that Russia used social media as a cyber-weapon in the conflict with Ukraine, which in turn contributed to the reform of the security and defence sector in Ukraine.

The Russia-Ukraine conflict demonstrates the growing influence of cyberspace and social media as a digital platform in the conflict landscape of the twenty-first century in support of conventional military action, taking into account that the annexation of Crimea and the continued conflict in Eastern Ukraine, namely in the

---

[1] See, for example, Fried and Polyakova (2018).

[2] This study does not seek to provide a deterministic point of view on technology, but rather to challenge traditional conceptions of war, namely regarding what counts as a weapon and what kind of harm a weapon could cause to be considered as one.

Donbass region, coincided with Russia's extensive use of social media for disinformation and cyber-attacks against Ukraine (Zeitzoff 2017; Danyk et al. 2017; Herrick 2016; Lange-Ionatamishvili and Svetoka 2015).

The present contribution is divided into three sections. In the first section, we introduce the debate on the definition of cyber-weapons. In the second section, we examine the opportunities and risks associated with the use of social media by the armed forces for strategic communications operations. Furthermore, we demonstrate that the hostile use of social media expands beyond strategic communications operations and that it can be used both at the operational and the tactical level. In the third section, we demonstrate how social media was used in the Russia-Ukraine conflict as a cyber-weapon and examine its implications.

2 Cyber-Weapons

Despite the undeniable room that emerging cyberspace-related threats and challenges is given on the post-Cold War international security agenda, cyberspace was not always considered a security issue. The term 'cybersecurity' was introduced in the early 1990s by computer scientists, referring to insecurities related to networked computers and to the need to protect data existing in computer systems as well as the computer systems themselves against unauthorised external intrusion (Hansen and Nissenbaum 2009).

However, political elites, the private sector, and traditional media quickly recognised the potential for the political and social effects of computer systems and their implications for security (Valeriano and Maness 2018; Hansen and Nissenbaum 2009). This securitisation[3] entails two core elements, one technical and one social. On the one hand, the technical element is related to the network character of computer systems. These systems control physical objects such as trains, pipelines, and electrical transformers. This means that in the event of a cyber-attack, these systems can be compromised, which in turn can hinder or prevent electrical or communication distribution, disrupt transportation systems, disable financial transactions, and consequently generate chaos (Hansen and Nissenbaum 2009). The social element, on the other hand, is related to the technological dependence of states and societies, which – considering the growing number of processes that rely on the digital domain – creates a perception of vulnerability (van der Meer 2015). Accordingly, the concept of cyberspace was expanded beyond a technical issue to be also recognised as a security issue.

Cyber-weapons had a similar evolution. Despite being used as military and intelligence tools since the 1990s, it was only at the end of the first decade of the

[3] For a detailed understanding of this securitisation process see, for example, Hansen and Nissenbaum (2009).

2000s – in 2007 and 2010 with Estonia[4] and Stuxnet[5] respectively – that the strategic capability of weaponised code and its potential to destabilise international relations and to be used as a tool to accomplish national strategic interests was recognised, producing an intense debate on the regulation of the acquisition and use of these type of weapons[6] (Stevens 2017).

Notwithstanding, as Stone (2013) notes, the ongoing debate on cyberspace in general exposes the precarious understanding of concepts that are routinely employed in the security and defence studies framework, such as weapons and, in this case, cyber-weapons. The term cyber-weapon is yet to be given a consensual definition. Thus far, it has been used as a catch-all term for different forms of the hostile use of cyberspace, resulting in a lack of analytical rigour. Furthermore, the absence of conventional physicality – meaning code only exists in infrastructures and is difficult to track and interdict – and the subjective process of establishing the nature and degree of what constitutes harm hamper the efforts towards a clear definition of what is considered a cyber-weapon (Stevens 2017).

Nevertheless, some efforts have been made to deepen the understanding of what constitutes a weapon in general and a cyber-weapon in particular. Traditional wisdom understands weapons as tools that cause or are designed to cause harm or damage, namely physical harm and damage (Meiches 2017; Bousquet et al. 2017; Stevens 2017; Rid and McBurney 2012). In addition, weapons are instruments that are used or designed to be used in order to enhance power and to alter will or desire (Meiches 2017; Bousquet et al. 2017). Therefore, weapons can be defined as tools that are used or designed to be used with the aim or the potential to cause physical, functional, or mental harm to structures, systems, and living things in order to enhance power (Meiches 2017; Bousquet et al. 2017; Rid and McBurney 2012). In this scenario, cyber-weapons are understood as a subset of weaponry; "a computer code that is used, or designed to be used, with the aim of threatening or causing physical, functional, or mental harm to structures, systems or living beings", in order to enhance power (Rid and McBurney 2012).

According to Rid and McBurney (2012), even though the line between what does and does not count as a cyber-weapon is subtle, it is important to identify it for several reasons: (1) security reasons – if a tool has no potential to do harm it is less dangerous; (2) political reasons – an unarmed intrusion is less urgent than an armed one; (3) legal reasons – recognising something as a weapon means that in principle it can be outlawed and its development, possession, and use may be punishable. These distinctions are important to develop appropriate and proportionate responses. Therefore, according to Rid and McBurney (2012), cyber-espionage, for instance, is

[4] For a better understanding see, for example, Boyte (2017).

[5] Stuxnet is a malicious computer worm, that targets SCADA (supervisory control and data acquisition systems) and, although neither the USA nor Israel openly admitted its use, it is believed that their use of Stuxnet was responsible for causing damage to the nuclear programme of Iran. For a better understanding see, for example, Lindsay (2013).

[6] For a better understanding see, for example, Eilstrup-Sangiovanni (2018).

not considered a weapon, although it is probably the most common form of cyber-attack.

Nevertheless, as was mentioned above, the core goal underlying the use of weapons is to enhance power (Meiches 2017; Bousquet et al. 2017; Rid and McBurney 2012). As the next section tries to demonstrate, evidence related to the hostile use of social media shows that physical destruction is not the only way to do harm or damage and to enhance power, at least in the digital domain – and especially within the military framework.

3 Social Media and the Armed Forces

3.1 Armed Forces and the Use of Social Media: Opportunities and Risks

The implications of the media landscape to war in general and to the armed forces in particular are not a novelty. Nevertheless, the new information environment and the increasing use and role of social media introduced new opportunities as well as new risks and challenges to the military (Ryan and Thompson 2017; Müür et al. 2016; Olsson et al. 2016; Hellman et al. 2016).

The use of social media within the military framework contains an inherent contradiction. The armed forces are considered a closed, formal, and secretive structure that communicates in a formal manner, thus contrasting with the informal, open, and emotional type of communication on social media. Notwithstanding, these platforms have been used by the military both at the national and the international level. At the national level, social media is usually used as a tool for recruitment as well as to engage with civil society in order to present and clarify the military's mission and activities. At the international level, social media creates new opportunities for the armed forces to share their message in areas of deployment, to boost their reputation, and to communicate their national strategic narrative in order to win the increasingly relevant battle for hearts and minds (Olsson et al. 2016; Hellman et al. 2016). Therefore, one could argue that social media offers an interesting platform through which the armed forces can interact with their audiences – not only to recruit, but also for strategic communication that seeks to improve transparency and legitimise actions (Golan and Ben-Ari 2018; Ryan and Thompson 2017; Hellman et al. 2016; Mirrlees 2015).

However, the global reach, open access, and speed of sharing information represent both the strengths and the weaknesses of social media. Moreover, the unpredictable and uncontrollable features of these digital platforms mean that control over the narrative can be lost, which can in turn endanger military personnel, military operations, and the fulfilment of strategic objectives. Risks concerning information have always existed in the defence sector, but with the advent of social media they can take on new forms. The distribution of information may expose soldiers through the dissemination of sensitive intelligence and/or may reconfigure public perceptions not only of political matters, but also with regard to defence activities (Ryan and Thompson 2017; Olsson et al. 2016; Hellman et al. 2016).

According to Ryan and Thompson (2017), the emergence of smartphones and the constant sharing of information on social media has increased the risk for military personnel to inadvertently breach security. Moreover, the exposition of sensitive information can be exploited by malicious actors as a source of intelligence, to uncover passwords, for social engineering, identity theft, physical interception, and blackmail (Ryan and Thompson 2017; Olsson et al. 2016; Hellman et al. 2016). Therefore, given the implications to security, the purpose of the mission, and the reputation of the armed forces and their activities, social media potentially is making the military vulnerable (Ryan and Thompson 2017; Olsson et al. 2016; Hellman et al. 2016). Consequently, to military operations, social media platforms such as Twitter, Facebook, and YouTube have become just as relevant as ammunition, troops, and air power, and are now being recognised as a tool and a contested battle space of the twenty-first century (Lange-Ionatamishvili and Svetoka 2015; Mirrlees 2015).

3.2 The Hostile Use of Social Media Beyond the Opportunities and Risks of Strategic Communications

The military use of social media extends beyond the opportunities and risks of strategic communications. Social media can have strategic value at the operational and tactical levels and can contribute to the deployment of several military activities, expanding the hostile use of social media (Herrick 2016). The hostile use of social media – which is never or almost never outlawed, thus making retaliation more complex – is a widespread phenomenon which concerns the abuse of social media platforms by governments, the private sector, organisations, and individuals in order to pursue political and military goals (Willemo 2019).

The conceptualisation of social media proposed by Nissen (2015, 40) demonstrates the expansion of these platforms beyond their use as an instrument for communication:

> Internet connected platforms and software used to collect, store, aggregate, share, process, discuss or deliver user-generated and general media content … can influence knowledge and perceptions and thereby directly or indirectly prompt behaviour as a result of social interaction within networks.

Contrary to this, Zeitzoff (2017) defines social media more narrowly as “a form of electronic communication and networking sites that allows users to follow and share content (text, pictures, videos, etc) and ideas within an online community”. For the purposes of this research, we found that the definition proposed by Nissen (2015) is more appropriate, considering that his conceptualisation of social media goes beyond its role as a tool for communication.

The recognition of the successful campaigns promoted by both Russia and the Islamic State have generated a shift in perception regarding the relevance of the information threat (Giles n.d.). The alleged Russian interference in the 2016 U.S. presidential elections and in European referendums and elections, such as Brexit and the French presidential elections, emerged as a game changer regarding the paradigm of social media, which, during the Arab Spring, had been perceived as a positive tool. The perception of social media has changed from being a positive tool that enabled social mobilisation during the Arab Spring to becoming a negative tool used for the proliferation of hate speech, recruitment for terrorist organisations, and disinformation campaigns, where facts are less influential than emotions for the formation of public and elite opinion (Roozenbeek and van der Linder 2018; Jankowski 2018; Bjola 2017).

Nevertheless, the hostile use of social media is not a new phenomenon. Long before the emergence of the Islamic State, Osama bin Laden had already recognised the opportunities of modern media and used the interconnectivity of the internet for subversion and information activities (Giles n.d.). In the early 2000s, terrorist organisations took advantage of social media to challenge the legitimacy and credibility of the multinational force led by the USA in Iraq and Afghanistan. Terrorist organisations promoted narratives on social media to influence the population of the contributing countries in order to have them question and oppose the presence of their respective countries in those regions. The central goal of these tactics was to change the centre of gravity from the physical to the cognitive domain through the proliferations of narratives and images on social media (Nissen 2015).

Social media has played a relevant role in several conflict situations since its introduction to the political spectrum. In the attack of Al-Shabaab in the Westgate Mall in Nairobi, the terrorist organisation used Twitter to live-tweet the attack. Islamic State used it to promote propaganda and to recruit, while in the Iranian Twitter Revolution between 2009 and 2010 social media was used to show the abuses carried out by the Iranian regime to the outside world. Finally, in 2010, social media played a relevant role in the development of the Arab Spring, contributing as a space for freedom of expression and enabling mass mobilisation (Nissen 2015).

Social media has become part of contemporary conflicts and politics and has been used by both state and non-state actors to create effects in the virtual as well as in the physical domain (Zeitzoff 2017; Danyk et al. 2017; Nissen 2015). According to Nissen (2015), social media has become a weapon of choice because of its easy access and use and the cost-effect gains, contributing to the efficiency of several military activities aiming to accomplish political and military objectives, such as intelligence, psychological operations, offensive and defensive cyber-operations, and command and control (C2).

Intelligence in this context refers to a group of activities that involve the collection, processing, integration, analysis, evaluation, and interpretation of data. In social media, intelligence means monitoring online activity to collect and aggregate data in order to generate knowledge in general and to support the process of targeting[7] in particular (Nissen 2015).

The proclivity to share posts creates an immense volume of data, making social media a fertile ground for intelligence. Through the aggregation and correlation of multiple sources of information, actors can assess the demographic size, organisational structure, areas of activity, and network reach of their target. Moreover, due to geo-location available on social media, malicious actors can locate and identify troop movements (Ryan and Thompson 2017; Marcellino et al. 2017). One example relates to the war on terrorism when, in 2015, a geo-tagged post on a fighter's social media account caught the attention of a U.S. Air Force unit, resulting in the launch of a bombing campaign on a building belonging to ISIS (Marcellino et al. 2017). Furthermore, these platforms provide the possibility to collect data online and in real time, depending on the speed of monitoring and analysis software, without requiring "boots on the ground" (Nissen 2015, 61–64). Therefore, social media is an attractive tool to monitor and document an adversary's activities (Herrick 2016).

Psychological operations refer to a group of military activities aimed to influence the perceptions, emotions, motives, reasoning, and behaviour of target audiences in favour of the objectives of the attacker. Psychological operations can be carried out overtly or covertly through actions such as deception, propaganda, and subversion in order to shape, inform, influence, manipulate, mislead, expose, coerce, deter, and mobilise (Nissen 2015). Social media offers a fertile ground for psychological operations, considering its technical features that allow the multiplication of information at high speed with little cost and a global reach. Moreover, its trust-based character, formed by a network of friends or like-minded members, makes the narrative more trustworthy and believable than the one presented by official media and/or governmental institutions (Lange-Ionatamishvili and Svetoka 2015).

State and non-state actors use psychological operations to inform and expose facts and actions that can challenge the credibility of adversaries and by means of manipulating information, such as the promotion of disinformation campaigns. The use of social media to promote psychological operations, namely through the manipulation of information, reveals a novel tendency in cyberspace, which stands in contrast with traditional forms of cyber-attacks. Such operations target societies with the specific goal to influence their believes and behaviours in order to sow discord and diminish their trust in institutions and governments (Kalpokas 2017; Prier 2017; Lange-Ionatamishvili and Svetoka 2015).

With regard to social media, state and non-state actors try to control and explore trend mechanisms in order to create distrust in public and private institutions to cause internal conflict. For this purpose, certain narratives are disseminated on these digital platforms and, by means of algorithm-driven mechanisms, create trends that almost instantaneously spread on a global scale (Prier 2017). Therefore, this dynamic allows a narrative to become viral, which turns it into a "low-cost, speedy

[7] In this context, targeting means to direct a disinformation campaign at a specific person or group (Roose 2018).

way of manipulating society's perceptions in order to cause disruptive behaviour in real life" (Lange-Ionatamishvili and Svetoka 2015, 105). In addition, the ambiguity associated with cyberspace complicates the identification of the source, preventing the planning of a response to these activities (Libicki 2017). In this sense, according to Prier (2017, 52), by creating "a point of injection for propaganda", social media "has become the nexus of information operations and cyberwarfare".

Cyber-operations on social media can be offensive or defensive. Offensive operations refer to several activities: that seek to deny access, disrupt, degrade, breach or destroy information; the hacking of passwords of personal accounts to expose content; and the intrusion to alter or insert content in social media profiles. Defensive operations imply the protection of social media platforms, accounts, and profiles by means of encryption, anti-tracking, and counter-narrative actions (Nissen 2015).

Command and Control (C2) refers to a group of activities that involve internal communication, information sharing, coordination, and synchronisation of actions on social media in order to facilitate the coordination and synchronisation of dispersed groups (Nissen 2015).

Although the effects of digital platforms on conflict scenarios are more indirect than conventional weapons systems, they can nevertheless influence the spread of information, both in scope and speed, which in turn can change the course of events (Zeitzoff 2017). According to Zeitzoff (2017), actors involved in conflict situations use social media not only to mobilise support, but also to actively shape the narratives about their group and their adversary as well as to support operational and tactical military actions. In line with this, the following section thus aims to demonstrate how Russia used social media in the Ukrainian conflict in order to destabilise the adversary at all levels, from the political to the military.

4 Social Media as a Cyber-Weapon? The Case of the Russia-Ukraine Conflict[8]

4.1 Russia-Ukraine Conflict[9]: Russian Military Operations on Social Media[10]

As demonstrated in the previous section, the use of social media, goes beyond the opportunities and risks of strategic communications. It can also be used to deploy military operations, for instance psychological operations, to gather intelligence, and to contribute to recruitment for terrorist organisations. Accordingly, social media has been used both in intrastate and interstate conflict scenarios, such as the Russia-Ukraine conflict (Herrick 2016). The use of cyberspace and social media in the Russia-Ukraine conflict demonstrates the growing influence of these digital platforms in the conflict landscape of the twenty-first century in support of conventional military action (Herrick 2016; Lange-Ionatamishvili and Svetoka 2015). According to Zeitzoff (2017), the annexation of Crimea and the continued conflict in Eastern Ukraine, namely in the Donbass region, coincided with Russia's extensive use of social media and cyberspace for disinformation and cyber-attacks against Ukraine. This assumption is shared by Danyk et al. (2017), who state that combat operations in Illovaysk and Debalcevo were preceded by increased activity in the information environment, which entailed an increase in spreading negative information about the Ukrainian authorities, government, and armed forces. Social media platforms were used by Russian military forces, intelligence agencies, and proxies to conduct several military operations (Herrick 2016). Cyberspace has strengthened Russia's power, on the one hand, by providing capabilities that enabled the realisation of its foreign policy objectives, either by facilitating the promotion of propaganda and coercion or to gather data. On the other hand, the difficulty to attribute attacks made available a cost-effective strategy (Ajir and Vailliant 2018; Mejias and Vokuev 2017).

In Ukraine, Russia used social media namely for recruitment, intelligence, and psychological operations.[11] Vkontakte and Odnoklassinski, Russian social

[8] This research does not aim to deepen the understanding of the underlying dynamics of the conflict, but to demonstrate the role of social media in it as a cyber-weapon.

[9] Social media was also used by the Ukrainian Armed Forces, government, and civil society to spread information or gain advantage (Herrick 2016). Nevertheless, we focus on the use of these platforms by Russia, considering its implications and role as a game changer in the paradigm of the information warfare landscape (Giles n.d; Herrick 2016; Lange-Ionatamishvili and Svetoka 2015).

[10] Although evidence demonstrates that Russia used both traditional and social media to mutually reinforce the narrative and address younger and older generations (Blank 2017; Lange-Ionatamishvili and Svetoka 2015), the goal of this chapter is to focus on the social media side, considering the novelty and challenges brought by these digital platforms.

[11] Russian psychological operations aimed at shaping different audiences: Ukrainians, Russians, and international audiences (Blank 2017). However, the goal of this chapter is to focus on Ukrainian perceptions, namely of the Ukrainian Armed Forces.

networks, were used to recruit agents for rebel groups, to gather data and information on the profiles of Ukrainian military personnel and civilians, and to deploy psychological operations at all levels, from the political to the military (Willemo 2019; Müür et al. 2016). The central objective of these operations was to destabilise the adversary – to confuse rather than convince – by cultivating fear, anxiety, and hate among the Russian-speaking population in Ukraine, spreading distrust between the Ukrainian government and the Ukrainian Armed Forces as well as between the Ukrainian government, the Ukrainian Armed Forces and civil society, and to demoralise and demobilise the Ukrainian Armed Forces (Giles n.d.; Cordy 2017; Sazonov et al. 2017; Müür et al. 2016; Lange-Ionatamishvili and Svetoka 2015).

The use of psychological campaigns[12] to mislead, manipulate, and intimidate is not a novel approach in the Kremlin's foreign policy. However, the internet and social media created new opportunities for the efficiency of old methods such as active measures[13] (Cordy 2017). According to Giles (n.d.), Russian information warfare is an ongoing process, which is constantly evolving, developing, and adapting by identifying success and reinforcing it, while abandoning failed attempts and moving on. In Ukraine, Russian psychological operations have been ongoing since the beginning of the 1990s, however, new opportunities were created by the advent of the internet and social media (Sazonov et al. 2017). Therefore, following the Euromaidan revolution,[14] a new set of psychological operations were put into effect (Cordy 2017; Blank 2017; Sazonov et al. 2017).

Social media was used by Russia in Ukraine to deploy intense psychological operations, namely by disseminating disinformation campaigns by means of trolls,[15] bots,[16] social engineering attacks[17] and impersonation[18] (Danyk et al. 2017). In this context, a 'troll factory' was reported in 2013, situated outside St. Petersburg and referred to as Internet Research Agency, where hundreds of paid bloggers – trolls – were encouraged to praise Putin and condemn the opposition. The crisis in Ukraine was the central battlefield for these trolls, who were asked to post comments that disparaged the Ukrainian president and depicted the alleged atrocities carried out by the Ukrainian Armed Forces (Mejias and Vokuev 2017). These bots and trolls were

---

[12] The focus on these operations relates to the fact that, in the approach to Russian information aggression, psychological operations are at the core of Ukrainian strategy.

[13] For a better understanding of the concept 'active measures' see, for example, Giles (2016); Ajir and Vailliant (2018).

[14] Euromaidan was a wave of civilian protests that began in November 2013 in Ukraine, sparked by the decision of the Ukrainian government to suspend the signing of the Association Agreement with the European Union in order to build a stronger partnership with Russia and the Eurasian Economic Union.

[15] In this context, a troll is a person who uses social media to sow chaos and discord.

[16] In this context, a bot is an automated account.

[17] In this context, social engineering refers to the manipulation and exploitation of human cognitive biases in order to exert influence (Willemo 2019).

[18] In this context, impersonation refers to the act of pretending to be another person, organisation, newspaper or website for the purpose of manipulation (Willemo 2019).

used to exploit human biases and vulnerabilities by amplifying certain narratives and manipulating information in order to destabilise, undermine cohesion, and fuel chaos. Moreover, such tactics also have the potential to undermine the situational awareness process, thus the manipulation of social media activity can compromise military decisions based on the analysis of activity on social media (Willemo 2019).

Such operations were used against all levels, from the political to the military. Their aim was to disseminate negative information about key authorities of the government by describing it as corrupt and fascist. Furthermore, they were used to diminish morale and promote the demobilisation on the armed forces with campaigns such as "Generals betrayers of Ukraine", "Hail to the Ukrainian Artillery", as well as to influence and sow confusion among the population (Danyk et al. 2017; Sazonov et al. 2017).

Russian psychological operations on social media towards the Ukrainian military were deployed by means of disinformation campaigns, which were based on collected data and the use of different myths and narratives. The Ukrainian Armed Forces were depicted as execution squads, Nazis, killers, and terrorists. Moreover, manipulated images were disseminated that exposed atrocities supposedly committed by the Ukrainian Armed Forces, including torture, civilians being used for organ trafficking, recruitment of child soldiers, the use of heavy weapons against civilians, and acts of cannibalism (Cordy 2017; Sazonov et al. 2017; Müür et al. 2016; Lange-Ionatamishvili and Svetoka 2015).

There are several examples of stories that were manipulated in order to discredit and demoralise the Ukrainian Armed Forces. In 2014, a photo of a war victim in Syria taken in 2013 was used as proof that Ukrainian soldiers had wounded a 3-year-old-boy in Sloviansk. The boy was allegedly tortured and crucified by the Ukrainian Armed Forces in a public square of Sloviansk in order to sow fear among the population (Mejias and Vokuev 2017; Lange-Ionatamishvili and Svetoka 2015). Furthermore, there is evidence of targeted messages to soldiers and officers that combined sensitive personal information and threats towards their families (Tsybulska 2019, cited in Willemo 2019). Additionally, the group most at risk of being manipulated within the armed forces were (and still are) young soldiers because they sometimes maintain the habits of civil life (Sazonov et al. 2017). In addition, rumours of the inhuman conditions of life in the Ukrainian Armed Forces were spread, describing that the military was being ruled by violence, chaos, and hunger, contributing to the desertion of hundreds and thousands of Ukrainian soldiers to join the Russian side (Sazonov et al. 2017).

According to Danyk et al. (2017), these operations aimed to discredit the Ukrainian authorities – especially the Ukrainian Armed Forces – resulted in disaffection and mistrust towards the Ukrainian military. Moreover, they created discord concerning the necessity of military actions and also damaged the morale of soldiers, which in turn encouraged them to desert. Hence, the fighting capacity of the armed forces was damaged, which in turn reduced the capability to respond to the aggression (Mölder and Sazonov 2018). According to Mölder and Sazonov (2018,

320), "the annexation of Crimea was a successfully conduct attack the Armed Forces were not prepared and surrendered without resistance".

4.2 *Trolls, Bots, and Reforms: The Implications of the Russian Information Operations to the Political, Security, and Defence Sector of Ukraine*

Assessing the implications of the military operations conducted by Russia on social media against Ukraine – particularly against the Ukrainian Armed Forces – regarding the offline attitudes and behaviour of military personnel is a demanding and probably impossible task. Thus, we propose a different approach by trying to identify the main steps that were taken to address these challenges within the Ukrainian political and military landscape. The first implication is the actual definition of measures to cope with these challenges, which is, at the minimum, public recognition of the problem itself.

In the political context, the events led to the creation of the Ministry of Information Policy of Ukraine in 2014. The creation of this new body was driven by the intention to counteract the information war conducted by Russia, which is reflected mainly in the mission and structure of this new body. The core mission of the Ministry of Information Policy of Ukraine is to safeguard and improve the Ukraine's information sovereignty and environment by means of the following measures:

- Improving citizen communications;
- Winning the information war in occupied and liberated Ukrainian territories;
- Promoting open governance and transparency on behalf of citizens;
- Protecting freedom of speech and the rights of journalists;
- Legislating media to European standards, requirements and recommendations.

Furthermore, the Ministry's structure – one minister and two deputies – also demonstrates its military-related dimension, considering that one of the deputies, who was nominated from the Ministry of Defence, has the responsibility to win the information war (Ministry of Information Policy of Ukraine n.d.).

In the political and military context, we highlight the approval of the Doctrine of Information Security of Ukraine in 2017 following the "Russian information aggression", whose purpose is to combat "the destructive information influence of the Russian Federation ... with the use of the latest information technologies" (Presidency of Ukraine 2017). One of several bodies responsible for the fulfilment of the objectives identified in this doctrine is the Ministry of Defence, which reinforces the military-related dimension of this threat. In addition, among the identified threats to Ukraine's national interests and security in the area of information are the challenges emanating from the operations that aimed at "undermining the defence capacity, demoralization of the staff of the Armed Forces of Ukraine and other military forces", which manifested in the Ukrainian preoccupation with the implications of information warfare conducted by Russia on the morale of the Ukrainian Armed Forces (Presidency of Ukraine 2017). Furthermore, within the military framework, the Ukrainian Armed Forces established a press office in order to support the strategic communications of military personnel. Additionally, a mobile group of psychologists and other professionals was formed, whose mission was to support soldiers by providing them with information that they lacked. Social media presence was also prioritised in order to bolster resilience towards information operations. Education, raising awareness, and implementing best practices were also recognised as vital components to deal with information operations (Willemo 2019).

Finally, in 2015, "the armed aggression of the Russian Federation against Ukraine" as well as "the Russian occupation of parts of Ukraine" and "the Russian incitement of conflict" resulted in the revision of the Military Doctrine and the National Security Strategy of Ukraine (Presidency of Ukraine 2015a, b). Therefore, "the Russian threat" once again provided an argument for fostering reform in Ukraine, particularly in the sector of defence and security. With regard to the Military Doctrine, information security was at the core of the reforms and even acknowledged as a game changer, among others, in relation to the nature of warfare. Hence, the improvement of the information policy in the military and the prevention of psychological operations conducted by foreign countries is an important objective identified of the doctrine (Presidency of Ukraine 2015a). In relation to the National Security Strategy, Russian information warfare was also perceived as a relevant threat to national security and thus an area were actions should be taken (Presidency of Ukraine 2015b).

5 Final Considerations

The novelty underlying cyberspace as a security issue is reflected, for instance, in the incipient conceptualisation of what counts as a cyber-weapon (Maness and Valeriano 2015; Hansen and Nissenbaum 2009; Eriksson and Giacomello 2006). The concept of cyber-weapon still lacks analytical debate, which results in a narrow understanding of what counts as the hostile use of cyberspace generally associated with the disruption of critical infrastructures (Stevens 2017; Lange-Ionatamishvili and Svetoka 2015). Nevertheless, the use of social media to conduct military operations, for instance psychological operations, has demonstrated that doing harm and damage in cyberspace can go beyond attacking machines through computer code (Lange-Ionatamishvili and Svetoka 2015). This reflection is important for two main reasons: on the one hand, there is an increased debate on how to regulate cyberspace, not only regarding the acquisition and use of weaponised code, but also in terms of social media (Stevens 2017). On the other hand, most of the operations analysed for this investigation that take place on social media are not outlawed, which makes retaliation even more complex.

In this context, this study analysed Russian's use of social media in the conflict with Ukraine in order to assess the potential of social media to be considered a cyber-weapon, thus going beyond attacks on critical infrastructures. In order to accomplish this goal, we used the conceptualisation of the term cyber-weapon proposed by Rid and McBurney (2012), which understands these weapons as computer code used to cause physical, functional or mental damage or harm to structures, systems or living beings. One opposition to this conceptualisation is the limitation to computer code, excluding other forms of the exploitation of cyberspace, such as the use of social media. As we tried to demonstrate, the exploitation of a social network beyond the use of computer code supported the military operations of Russia on the ground in Ukraine with political and security implications (Zeitzoff 2017; Danyk et al. 2017; Herrick 2016; Lange-Ionatamishvili and Svetoka 2015). We support, however, the claim made by Rid and McBurney (2012) regarding the importance of delimiting what does and does not count as a cyber-weapon for security, political, and legal reasons.

Nevertheless, as this paper tries to demonstrate, the use of social media in the Russia-Ukraine conflict contributed to cause damage to Ukrainian structures, systems, and living beings, namely to the Ukrainian Armed Forces. According to Danyk et al. (2017), those operations resulted in disaffection and mistrust towards the military, and in discord concerning the necessity of military actions. Furthermore, they damaged soldiers' morale and consequently encouraged them to desert. Hence, the fighting capacity of the armed forces was diminished, which in turn reduced the capability to respond to the aggression (Mölder and Sazonov 2018). Moreover, on the one hand, assessing the causality of social media use to manipulate attitudes and behaviours of military personnel may be challenging and difficult to prove. Yet, on the other hand, Russia's information aggression was used to officially justify the creation and mission of the Ministry of Information Policy of Ukraine in 2014, the approval of the Doctrine of Information Security of Ukraine in 2017, and the revision of the Ukrainian Military Doctrine as well as the National Security Strategy of Ukraine in 2015. This, at the minimum, is a reflection of the Ukrainian perception regarding the damage caused by Russia.

Based on these implications, we conclude that Russia used social media as a cyber-weapon in the conflict with Ukraine. In this regard it is important to consider not only the perceived damage those actions caused in Ukraine, but also to take into account the role they played in the reform of Ukraine's security and defence sector.

References

Ajir M, Vailliant B (2018) Russian information warfare: implications for deterrence theory. Strat Stud Q 12(3):70–89

Bjola C (2017) Propaganda in the digital age. Glob Aff 3(3):189–191

Blank S (2017) Cyber war and information war à la Russe. In: Perkovich G, Levite AE (eds) Understanding cyber conflict: 14 analogies. Georgetown University Press, Washington, DC, pp 81–98

Bousquet A, Grove J, Shah N (2017) Becoming weapon: an opening call to arms. Crit Stud Secur 5(1):1–8

Boyte KJ (2017) A comparative analysis of the cyberattacks against Estonia, the United States, and Ukraine: exemplifying the evolution of internet-supported warfare. Int J Cyber Warfare Terror 7(2):54–69

Cordy J (2017) The social media revolution: political and security implications. [Online]. NATO Parliamentary Assembly. (Viewed 20 December 2019). Available from: https://www.nato-pa.int/download-file?filename=sites/default/files/2017-11/2017%20-%20158%20CDSDG%2017%20E%20bis%20-%20SOCIAL%20MEDIA%20REVOLUTION%20-%20CORDY%20REPORT.pdf

Danyk Y, Maliarchuk T, Briggs C (2017) Hybrid war: high-tech, information and cyber conflicts. Connect Q J 16(2):5–24

Eilstrup-Sangiovanni M (2018) Why the world needs an international cyberwar convention. Philos Technol 31:379–407

Eriksson J, Giacomello G (2006) The information revolution, security, and international relations: (IR) relevant theory? Int Polit Sci Rev 27(3):221–244

Fried D and Polyakova A (2018) Democratic defense against disinformation. Atlantic Council

Giles K (2016) Handbook of Russian information warfare. [Online]. Rome: NATO Defense College. (Viewed 20 December 2019). Available from: http://www.ndc.nato.int/news/news.php?icode=995

Giles K (n.d.) The next phase of Russian information warfare. [Online]. NATO Strategic Communications Centre of Excellence. (Viewed 20 December 2019). Available from: https://www.stratcomcoe.org/next-phase-russian-information-warfare-keir-giles

Golan O, Ben-Ari E (2018) Armed forces, cyberspace, and global images: the official website of the Israeli defense forces 2007-2015. Armed Forces Soc 44(2):280–300

Hansen L, Nissenbaum H (2009) Digital disaster, cyber security, and the Copenhagen school. Int Stud Q 53:1155–1175

Hellman M, Olsson E-K, Wagnsson C (2016) EU armed forces' use of social media in areas of deployment. Media Commun 4(1):51–62

Herrick D (2016) The social side of 'cyber power'? Social media and cyber operations. [Online]. International Conference on Cyber Conflict, NATO CCD COE Publications. (Viewed 20 December 2019). Available from: https://ccdcoe.org/uploads/2018/10/Art-07-The-Social-Side-of-Cyber-Power.-Social-Media-and-Cyber-Operations.pdf

Jankowski NW (2018) Researching fake news: a selective examination of empirical studies. Javnost Public 25(1–2):248–255

Kalpokas I (2017) Information warfare on social media: a brand management perspective. Baltic J Law Polit 10:35–62

Lange-Ionatamishvili E, Svetoka S (2015) Strategic communications and social media in the Russia Ukraine conflict. In: Geers K (ed) Cyber war in perspective: Russian aggression against Ukraine. NATO CCD COE Publications, Tallin, pp 103–111

Libicki MC (2017) The convergence of information warfare. Strat Stud Q 11(1):49–55

Lindsay JR (2013) Stuxnet and the limits of cyber warfare. Secur Stud 22:365–404

Maness RC, Valeriano B (2015) The impact of cyber conflict on international interactions. Armed Forces Soc 42(2):1–23

Marcellino W, Smith ML, Paul C, Skrabala L (2017) Monitoring social media: lessons for future Department of Defense social media analysis in support of information operations. [Online]. Rand Corporation. (Viewed 20 December 2019). Available from: https://www.rand.org/pubs/research_reports/RR1742.html

Meiches B (2017) Weapons, desire, and the making of war. Crit Stud Secur 5(1):1–19

Mejias UA, Vokuev NE (2017) Disinformation and the media: the case of Russia and Ukraine. Media Cult Soc 39(7):1027–1042

Ministry of Information Policy of Ukraine (n.d.) Improving Ukraine's information environment. [Online]. Ministry of Information. (Viewed 20 December 2019). Available from: https://mip.gov.ua/files/Presentation/en_pres.pdf

Mirrlees T (2015) The Canadian Armed Forces "YouTube war": a cross-border military-social media complex. Glob Media J 8(1):71–93
Mölder H, Sazonov V (2018) Information warfare as the Hobbesian concept of modern times – the principles, techniques, and tools of Russian information operations in the Donbass. J Slavic Mil Stud 31(3):308–328
Müür K, Mölder H, Sazonov V, Pruulmann-Vengerfeldt P (2016) Russian information operations against the Ukrainian state and Defence Forces: April–December 2014 in online news. J Baltic Secur 2(1):28–71
Nissen TE (2015) The weaponization of social media: characteristics of contemporary conflicts. [Online]. Royal Danish Defence College, Copenhagen. (Viewed 20 December 2019). Available from: http://www.fak.dk/en/publications/Documents/The%20Weaponization%20of%20Social%20Media.pdf
Olsson E-K, Deverell E, Wagnsson C, Hellman M (2016) EU armed forces and social media: convergence or divergence? Def Stud 16(2):97–117
Presidency of Ukraine (2015a) Military doctrine of Ukraine. [Online]. Legislation of Ukraine. (Viewed 20 December 2019). Available from: https://zakon.rada.gov.ua/laws/show/555/2015#n17
Presidency of Ukraine (2015b) National security strategy of Ukraine. [Online]. Legislation of Ukraine. (Viewed 20 December 2019). Available from: https://zakon.rada.gov.ua/laws/show/287/2015
Presidency of Ukraine (2017) Doctrine of information security of Ukraine. [Online]. Council of Europe. (Viewed 20 December 2019). Available from: https://rm.coe.int/doctrine-of-information-security-of-ukraine-developments-in-member-sta/168073e052
Prier J (2017) Commanding the trend: social media as information warfare. Strat Stud Q 11(4):50–85
Rid T, McBurney P (2012) Cyber-weapons. Rusi J 157(1):6–13
Roose K (2018) Social media's forever war. [Online]. The New Work Times (Viewed 20 December 2019). Available from: https://www.nytimes.com/2018/12/17/technology/social-media-russia-interference.html
Roozenbeek J, van der Linder S (2018) The fake news game: actively inoculating against the risk of misinformation. J Risk Res 22(5):1–11
Ryan M, Thompson M (2017) Social media in the military: opportunities, perils and a safe middle path. In: Watola DJ, Macintyre A (eds) Technology and leadership: international perspectives. Canadian Defence Academy Press, Ontario, pp 217–225
Sazonov V, Müür K, Kopõtin I (2017) Methods and tools of Russian information operations used against Ukrainian armed forces: the assessments of Ukrainian experts. ENDC Occas Pap 6:52–66
Stevens T (2017) Cyberweapons: an emerging global governance architecture. [Online]. Palgrave Communications. (Viewed 20 December 2019). Available from: https://www.nature.com/articles/palcomms2016102.pdf
Stone J (2013) Cyber war will take place! J Strateg Stud 36(1):101–108
Valeriano B, Maness RC (2018) International relations theory and cyber security: threats, conflicts, and ethics in an emergent domain. In: Brown C, Eckersley R (eds) The Oxford handbook of international political theory. Oxford University Press, Oxford, pp 259–272
van der Meer S (2015) Enhancing international cyber security: a key role for diplomacy. Secur Hum Rights 26:193–205
Willemo J (2019) Trends and developments in the malicious use of social media. [Online]. NATO Strategic Communications Centre of Excellence. (Viewed 20 December 2019). Available from: https://www.stratcomcoe.org/trends-and-developments-malicious-use-social-media
Zeitzoff T (2017) How social media is changing conflict. J Confl Resolut 61(9):1970–1991

Misinformation and Disinformation in Social Media as the Pulse of Finnish National Security

Teija Norri-Sederholm, Elisa Norvanto, Karoliina Talvitie-Lamberg, and Aki-Mauri Huhtinen

Abstract Social media is becoming more and more of a security threat. Dissatisfaction with the content and quality of the information flow is increasing not only at a national level, but also at the level of people's everyday lives. Social media is one of the key channels for distributing misinformation and disinformation and has also become a key instrument for influencing political activity in particular. We define misinformation as shared information which is unintentionally false, whereas disinformation refers to false information which is purposefully shared for systematic informational influencing as well as for propaganda. The post-Cold War age has created a new global power order by using information – which is increasingly shared through social media – for political purposes. Small countries like Finland have become more and more dependent on the global information flow, while at the same time increasingly being subjected to the proliferation of misinformation and disinformation. Hence, social media has become an ever-more crucial factor in terms of national security threats. At the same time, however, it is also a potential platform for creating (generalised) trust in national security by means of sharing correct information among citizens. This study focuses on the flow of misinformation and disinformation on social media in relation to armed forces and national security. In this contribution, we also address issues related to the role of gener-

T. Norri-Sederholm (✉) · A.-M. Huhtinen
Department of Leadership and Military Pedagogy, National Defence University, Helsinki, Finland
e-mail: teija.norri-sederholm@mil.fi; aki.huhtinen@mil.fi

E. Norvanto
Finnish Defence Command, Helsinki, Finland
e-mail: elisa.norvanto@mil.fi

K. Talvitie-Lamberg
Faculty of Information Technology and Communication Sciences, Tampere University, Tampere, Finland
e-mail: karoliina.talvitie-lamberg@tuni.fi

E. Moehlecke de Baseggio et al. (eds.), *Social Media and the Armed Forces*, Advanced Sciences and Technologies for Security Applications,
https://doi.org/10.1007/978-3-030-47511-6_12

alised trust for psychological resilience and explore the European Union's role in countering disinformation.

Keywords Disinformation · Misinformation · Armed forces · Social media · Society · Generalised trust

1 Introduction

Digitalisation and social media as new opinion builders have changed the speed at which information travels, the way information is produced, and the way people are connected across national borders. The changing domain of the information space as well as the media landscape have majorly affected today's security environment (Kofman et al. 2017). The use of social media has also exposed Western countries to a new way of informational influencing. Social media is a key player in the support of authoritarianism taking media space from traditional journalism. It can be used to fuel tensions between different groups, cause mistrust towards governmental institutions and decision-makers as well as affect political discourse and even parliamentary elections. The exposure of citizens to large-scale disinformation, including misleading or outright false information, is a major challenge for societies and armed forces across the globe (Singer and Brooking 2018).

Informational influencing activities have malicious intentions, which aim at affecting the perceptions, behaviour, and decisions of the target groups. The different techniques are geared towards exploiting shortcuts in our thinking, either by learning about us through the data we share or by applying 'nudges' that manipulate our cognitive biases (Pamment and Agardh-Twetman 2018, 5). In countries like Finland, where the peacetime Defence Forces are based on the conscript system, it is also crucial to understand the aspect of young people and social media. Social media is a core part of young people's everyday reality. Thus, youth are constant targets of political, commercial, and social information intended to influence their opinions and actions. Moreover, they are exposed to hostile players, adversaries or trolls using social media for their own ends. Especially young people feel that social media is an integral part of their reality. Singer and Brooking (2018, 139) state that, since the "world has come to be ruled by the whims of virality and the attention economy, plenty of people seek to cheat their way to fame and influence. Plenty more happily sell them the tools to do so."

Social media has, in fact, become a security threat, given that its "decentralised technology … allows any individual to ignite the cycle of violence" (Singer and Brooking 2018, 13). Distributing misinformation and disinformation as well as exerting influence at all levels and in all sectors of society is the new global power order in the rhizome of the internet. Spreading misinformation and disinformation is a growing worldwide phenomenon, a new type of hybrid warfare, which will be used by more and more state players trying to achieve their political goals and subverting political processes in other countries. Moreover, social media makes sure that we can always find others who share our ideas, however bad and terrible they may be.

> On Twitter, popularity is a function of followers, 'likes', and 'retweets' ... On Google, popularity is a function of hyperlinks and keywords; the better trafficked and more relevant a particular website, the higher it ranks in Google search results. On Facebook, popularity is determined by 'likes' from friends and the updates that you choose to share. The intent is to keep users emotionally grafted to the network. Bombard your friends with silly, salacious news stories and you'll find yourself receiving less and less attention; describe a big personal moment (a wedding engagement or professional milestone) and you may dominate your local social network for days (Singer and Brooking 2018, 139).

Military organisations are also becoming more than mere messengers of the governmental strategy narrative – a trend that is not without its challenges. The demarcation line between politics and military actions has become blurred. The more politics are being performed and acted out through social media, the more armed forces are becoming involved in social media. However, armed forces will also see advantages from using social media, which increasingly is becoming the space where geopolitics are being shaped (Mangat 2018). "For the government, all communication is strategic information-sharing to achieve particular results for the state" (ibid., 16). Armed forces join this resource competition in the democratic Western societies.

The security of our society faces a variety of threats, both military and non-military. These security threats are intertwined and difficult to predict, and thus come with little warning. Changes in the operating environment and the threat scenarios that mis- and disinformation bring to armed forces have therefore given rise to a need to examine security as broadly as possible.

The purpose of this contribution is to describe the connection of misinformation and disinformation in ambivalent social media to armed forces and national security, as well as to understand the mechanisms and spreading of information on social media platforms. In order to do this, we describe the challenges and possibilities of social media for armed forces that are based on conscription. As a framework for our contribution, we highlight the role of the European Union (EU) in countering disinformation – given that the EU plays an important role in creating the rules and common behaviour systems regarding social media and security – as well as the role of generalised trust for psychological resilience.

2 Misinformation and Disinformation

Misinformation and disinformation have rapidly evolved as key concepts in public debate when talking about online social media participation. Participation in social media plays an important role for societal and political engagement among young people. Particularly the sharing of information like news content with peers is an important form of political commitment. Thus, given that social media has become the primary media through which young people develop their political identities (Woolley and Howard 2017) and understanding of societal issues, the reliability of the shared information becomes crucial. The World Economic Forum even identified the spread of misinformation online as being among the top ten perils to society (World Economic Forum 2014).

The public concern over misinformation and disinformation has emerged due to cases such as the 2016 U.S. presidential election and "fake news" (Bakir and McStay 2018; Guess et al. 2018; Woolley and Howard 2017), the 2016 UK Brexit referendum (Woolley and Howard 2017) and the events in Ukraine (Sanovich 2017; Smith 2015), as well as Jade Helm 15, a military exercise in the U.S. perceived on the internet as the beginning of a new civil war (Del Vicario et al. 2016), or misinformation about the Ebola epidemic among healthcare workers (Del Vicario et al. 2016). In all these examples, social media was used as a platform for spreading propaganda, setting political agendas, and engaging citizens politically based on sharing mis- and disinformation on a global scale.

The disinformation phenomenon thus has become a global concern. Public interest particularly arose based on the reports released by the U.S. Senate's Intelligence Committee, which revealed attempts by Russia's Internet Research Agency (IRA) – a Russian social-media propaganda machine – to influence the 2016 presidential election via disinformation efforts. The Committee showed that the IRA attempted to influence the political opinion forming of U.S. citizens through trolling, reaching its targets on YouTube, Instagram, Facebook, and Twitter. The influence campaign involved 187 million 'engagements', among which was the use of memes trying to change people's political mindsets. Memes have actually become powerful communicational means of cultural influence that may even change people's values and behaviour (Lyngaas 2018; Thompson and Lapowsky 2018). According to the report, "over the past five years, disinformation has evolved from a nuisance into high-stakes information war" (DiResta et al. 2018, 100).

In order to understand these new forms of information sharing and the fundamental threats they pose to societies, a more precise definition of the concepts of information, and particularly of dis- and misinformation, would be beneficial. In public debate, the terms *misinformation* and *disinformation* are generally used when referring to informational content which is incorrect or untrue. Among information researchers, information is defined as "meaningful data, which might or might not be true" (Fetzer 2004, 224) and as "something that has representational content" (Fallis 2014, 137), regardless of its truth value. Conversely, information is also generally considered as something that is, in fact, 'true' (Floridi 2011).

Pragmatic understanding defines misinformation as information whose representational content is inaccurate (Fallis 2015; Karlova and Fisher 2013; Søe 2018). Misinformation also refers to "well-formed and meaningful data (i.e., semantic content) that is false", mistaken or misleading (Floridi 2011, 260). However, the content of the information is not as important as the agency and the intent behind sharing the information, which is what makes all the difference between misinformation and disinformation. While misinformation describes false or inaccurate information that is *not* purposefully intended to deceive, nor necessarily wilfully distributed (Fetzer 2003; Kumar and Geethakumari 2014, 3), disinformation is intended to deceive and confuse by being "purposefully conveyed to mislead the receiver into believing that it is information" which is true (Floridi 2011, 260; see also Søe 2018). It is important to note that disseminating incomplete information with the aim of deliberately deceiving others about the truth may also be a form of disinformation (Fallis 2015; Kumar and Geethakumari 2014).

Since disinformation is the intentional "distribution, assertion, or dissemination of false, mistaken, or misleading information" (Fetzer 2003, 228), the agency behind distributing the information becomes essential. The difference between misinformation and disinformation is connected to the question of agency in information sharing. As Søe (2018) points out, the difference between mis- and disinformation lies in the intentionality of misleading. When considering the central role online environments play in the political and societal engagement of young people, individual agency in forming the information becomes crucial. Sharing misinformation is an unintended process, however, the distributor of such information may also become an important node in disinformation networks. Simply by innocently sharing content that seems interesting, the individual may unknowingly act as a contributor in a planned deceit process of disinformation (Kumar and Geethakumari 2014, 14). It is catchiness, not truthfulness, that often motivates information diffusion on social media (Kumar and Geethakumari 2014; Ratkiewicz et al. 2010). This is connected to the understanding of information as a communicational process, where the context and the purpose of the sharer of information defines the final value and content of the information itself.

In this regard, the true/false dichotomy for distinguishing between information and mis- and disinformation seems insufficient. This becomes particularly evident when examining the information sharing habits of young people. In a study by Chen et al. (2015) on how young people share misinformation, over 60% of respondents shared misinformation, while 85% indicated that they might share misinformation in the future. The top reasons were related to the information's perceived characteristics as well as to self-expression and socialising. The reasons given for sharing misinformation relating to information characteristics included that it offered a good topic of conversation, that it was interesting, and that it was "eye-catching" (Chen et al. 2015, 587). The social reasons given for sharing were that it enhanced interpersonal relations and interaction with people, and that sharing was understood as a culture (ibid.). The accuracy of information and the authority of the information sources were among the least important reasons for sharing, as information shared by peers is understood to be reliable enough, which is also mentioned in several other studies (e.g., Kim et al. 2014). This is known as an echo chamber effect, where likeminded peers share information that follows their (shared) ideologies (Del Vicario et al. 2016; Dubois and Blank 2018).

The fact that information might be transformed into mis- and disinformation in online social interaction and used for propaganda naturally challenges top-down, organisational communication and information sharing when framing societal issues. One such issue is young people's understanding of safety and security. Tackling the circulation of mis- and disinformation is not merely a matter of detecting false or inaccurate information, but demands a more holistic understanding of why information, for example on safety threats, becomes commonly distributed and whose agenda it may serve.

3 Armed Forces and Social Media

The following chapter is based on an interview with a captain working in the Public Information Division in Defence Command, who has 15 years of expertise in military strategic communication and information warfare. He holds a Master's degree in Political Science and is also a film producer. This chapter is a summary of the main points of the interview conducted in May 2019. The duration of the interview is 85 min, its verbatim transcription amounting to a total of nine pages (Times New Roman 12 pt., single spacing). This interview, through the voice of a subject matter expert, shows that in Western armed forces, strategic communication still has not established itself, and that experts of this field are rare in the armed forces.[1]

Generally, according to the interviewee, there are some basic aspects to social media and armed forces in Western countries. Under normal circumstances, social media is essentially a tool for communication. At the same time, however, it poses a threat to armed forces: social media can be used as a weapon against them because information can be weaponised and used for influencing. In essence, social media is a collection of channels, messages, and opinions in which armed forces need to actively participate.

The interviewee notes that information has been used as a weapon for centuries. Messages and propaganda have been spread aiming to mislead, to frighten or to demonstrate capability to one's people or to the adversary, or just to deliver a political message. Therefore, the weaponization of social media is quite natural. As a global platform, and thus a huge agora where everyone can hear everyone's shout, it is actually a rather effective gun carriage. Operation Valhalla in Iraq in 2006 is an excellent example of using social media to deliver disinformation, resulting in the withdrawal of hundreds of U.S. Special Forces soldiers from combat. Two Special Forces Units had defeated part of the Jaish Al-Mahdi terror group in a successful engagement in a mosque. However, they left the bodies there. Afterwards, other Jaish Al-Mahdi fighters came, removed the weapons and framed the situation to make it look as though the dead had been praying. They then recorded a video depicting the slaughter of praying Muslims by the U.S. Armed Forces. This caused a media commotion, resulting in an investigation during which Special Forces were withdrawn from combat. This information operation created several problems.

[1] All statements in the following chapter are based on the interview. For an easier reading flow, in-text citations were omitted, except in case of direct quotes.

Many Muslims were enraged by the alleged slaughter, which only increased their willingness to fight. Moreover, several hundred Special Forces fighters were eliminated from the battlefield with just one video. Fortunately, the Special Forces soldiers had a Combat Camera Team with them and were in the end able to tell the truth.

From a military aspect, social media is a complex environment, and several dimensions need to be taken into account. There are many examples of how intelligence agencies have used the content in social media for targeting or – vice versa – how, by tracking and doing detective work, operators like Bellincat were able to reveal cases such as Skripal. Social media is a vast information bank for military purposes. According to the interviewee, nowadays armed forces understand the facts related to information warfare. "From an operative aspect, it is possible to win or lose battles without firing a single shot. Everything happens in the information environment" (Interview with subject matter expert).[2]

In Finland, such tactics used to be called *psychological operations*. Nowadays, they are referred to as *information defence*, which also includes the offensive aspect. It is one form of manoeuvre. Another dimension is the Command and Control system (C2). Armed forces should be agile both in the information environment and in social media. However, in reality, all systems must be well protected from cyberattacks and cyber influencing. This poses quite a challenge when considering usability and open platforms. Armed forces need to choose and use a wide variety of tools, such as applications, networks, and accounts, and have control over them within a large organisation and its bureaucracy. Thus, the agile and rapid nature of social media is a challenge for armed forces, including the Finnish Defence Forces, for several reasons. Firstly, a public safety authority always has a social and legislative responsibility. Secondly, there is a lack of agility in organisation culture. Thirdly, according to the interviewee, the tools and systems in cyberspace become rigid and slow by the necessity to protect them.

Despite these challenges, the interviewee also emphasises that under normal circumstances, social media is the place for armed forces to share correct information with citizens. By using different types of platforms (YouTube, Instagram, Snapchat, Spotify, and Facebook), it is possible to reach a wide audience. However, being able to release current information of good quality on six or seven different channels requires considerable resources:

> In countries like Finland, where defence is based on general conscription instead of a professional army, it is vital to keep up the conversation in society and thus take care of society's and citizens' willingness to have a conscription army. At the call-up process, 18-year-old youngsters make a choice between military service and non-military service. By delivering mis- and disinformation in social media, it is possible to influence adolescents' mindsets and reduce their will to defend their country (Interview with subject matter expert).

The interviewee further highlights that regardless of whether social media is being used as a means of communication in peacetime conditions or as a weapon in

---

[2]All direct quotes taken from the interview were translated into English by the authors of this contribution.

wartime conditions, armed forces need to be ready at all times to react to the intentional spreading of mis- and disinformation for ill intent. However, analysing the information environment is challenging. Traditional ways of influencing, such as framing, are still in use, and social media offers excellent platforms for them. The key to successfully spread mis- and disinformation is to have a reliable platform where people spend time daily and, unsuspecting of anything bad, share weaponised information. There need only be one or two players to start a fire or add fuel to one that is already burning. Alternatively, the players could be bots and it actually might be artificial intelligence we are fighting. Deepfake videos and memes have become a common way to influence and manipulate people's thinking – even for people with good media literacy skills it can be difficult to recognise the underlying purposes.

"Social media has become a tool for smaller and more agile troops to do battle at the same level as countries with powerful armies", as the interviewee points out. Indeed, the world has changed, and there are examples where countries like the United States or Israel won the war on the battlefield – only to realise that the battle is not over, but is still ongoing in the information environment. According to the interviewee, there are countries that utilise weaponised information in order to deliver disinformation to destabilise democratic societies and exploit vulnerabilities. However, democratic countries, such as Finland, cannot weaponise social media, as it is simply not acceptable to use these kinds of mechanisms and acts. Despite the fact that the enemy does not follow the rules of the game, democratic countries should uphold them. The interviewee emphasises that therefore, the aim of the Finnish Defence Forces is to be open, honest, and credible; to share ethically and morally strong stories, which are easy to tell and which citizens can believe. It is crucial to remember that even after building up a reputation for 20 years, it can be lost in seconds. Fighting on the information battlefield thus is a balancing act of having the necessary influence, while, at the same time, being ethical and credible.

In the following, we argue how generalised trust towards society should be understood as an effective tool against the sharing of mis- and disinformation in the era of information warfare. Strategically, this trust is generated intra-nationally as well as extra-nationally, for example by means of EU initiatives.

4 Generalised Trust and Resistance to Misinformation and Disinformation

One of the key factors affecting the subjective experience of safety and security is the trust citizens have in institutions that provide security (Limnéll and Rantapelkonen 2017). The public perception of the security authorities is based on the information to which citizens are exposed as well as on the criteria by which the public evaluate the actual performance, accountability, and transparency of the authorities (Blind 2006; Boda and Medve-Balint 2017; Kasher 2003). Nevertheless, in the complex information environment, targeted information campaigns can have an impact on how trustworthy the national security and safety bodies are perceived by citizens (Pamment et al. 2018). In liberal Western society, opinion formation takes place in the public sphere (Gripsrud et al. 2010), where different narratives can strongly influence the opinion of the masses and undermine the authority and trustworthiness of the government and other public institutions (City of Helsinki 2018; Håkansson and Witmer 2015; Pamment et al. 2018). Therefore, it is important to consider why citizens place their trust in national authorities and how information influencing activities can exploit this.

A fundamental feature of liberal democracies with a free mediasphere is trust and a high level of confidence between the different players within society. Nevertheless, a commitment to upholding law principles opens the door for hostile players who wish to exploit legislation to their advantage. Disinformation and misinformation can pose a threat to national security if they lead to an erosion of general trust (Hybrid CoE 2019). Many countries around the world, including EU member states, are taking action to combat misinformation and disinformation. Countermeasures range from government task forces, media literacy campaigns, public-radio podcasts, and public handbooks to internet shutdowns as well as the implementation of new bills and laws and regulations targeting hate speech, the spread of fake news, and foreign propaganda (Funke and Flamini 2020; The Law Library of Congress 2019). As the central characteristic that distinguishes disinformation from misinformation relates to the agency and intention of the propagator (Fetzer 2003; Fallis 2015), policymakers face a daunting challenge in trying to structure potential regulations to counter both. The inherent challenge relates to the compatibility of the existing human rights framework, such as freedom of expression and effective mitigating measures (Lotti 2018). The danger lies in the fact that initiatives aimed at countering specific problems of disinformation or misinformation – either by accident or design – enable public or private authorities to restrict free speech (Gutierrez 2018.)

Public measures regulating online content, for instance censorship, pose a challenge in liberal democracies and hence may not be the most effective measure to combat dis- and misinformation. Consequently, besides the restrictive measures to control online content, the need to build up citizens' resistance to disinformation has been identified (Palmertz 2016; West 2017). In this regard, measures to promote media and information literacy can help citizens navigate the social media environment. In fact, research suggests that building social resilience is also an essential contributor to resistance towards disinformation and misinformation (City of Helsinki 2018; EEAS 2019a; West 2017). Social resilience refers to a society's ability to cope with uncertainty and to "recover from shocks and emergencies" (Giacometti et al. 2018, 5). Research suggests that there is a positive correlation between societies with high social or generalised trust and social resilience (ibid., 6).

It can thus be assumed that societies with a higher public trust, such as those in the Nordic countries, are perceived to be more resistant to information influencing, as generalised trust promotes a sense of security and facilitates cooperation and interaction among citizens (e.g., Bäck and Kestilä-Kekkonen 2019; City of Helsinki 2018; Committee on Foreign Relations 2018; Giacometti et al. 2018; Pamment et al. 2018). For example, according to the Finnish concept of comprehensive security, the role of psychological resilience has been underlined as a fundamental factor substantiating the security of Finnish society (Security Strategy for Society 2017). Citizens' trust in each other and in official institutions, also known as 'generalised trust' (Rothstein and Stolle 2008), has a major impact not only on society's susceptibility to disinformation, but also on its resilience in the event of disruptions after the realisation of a threat (City of Helsinki 2018). In this context, building trust in the government and other policy-implementing bodies, such as the police, the military or the courts, is necessary to maintain social order and avoid serious clashes.

Trust has been discussed in many contexts relating to civic culture (e.g., Uslaner and Brown 2005), democracy (e.g., Offe 2000), and, more recently, media studies (e.g. Bennett and Livingston 2018). Generalised trust is considered to be one of the most important societal factors that make democracy work (Nannestad 2008; Putnam 1993; Uslaner 2002). Moreover, it is an important precondition for a stable and peaceful society where people tend to cooperate rather than defect (Newton 2001; Putnam 1993; Rothstein and Stolle 2003; Zak and Knack 2001). It is generalised trust that creates the feeling among citizens that the basic institutions and rules of law in democratic countries "can be relied upon to maintain a shared public good, to legitimate political opposition, smooth transitions of power and extend the protections of the rule of the law equally to all" (Abramson 2017, 5). In fact, generalised trust extends beyond the boundaries of face-to-face interaction and incorporates people who do not know each other personally (Uslaner 2002). Unlike society-centred approaches to generalised trust (Fukuyama 2001; Putnam 1993), the institutional approach suggests that state institutions facilitate the development and creation of generalised trust, thus providing a beneficial space in order to encourage trust and reciprocity (Levi 1998; Levi and Stoker 2000; Rothstein and Stolle 2008; Tarrow 1996). If citizens perceive government institutions, such as the Finnish Defence Forces, to be well-functioning and fair, they are likely to consider society to be safer and more secure as well as believe that most people have reason to behave honestly and, consequently, that most people can be trusted (Berg and Johansson 2016; Newton 2007; Rothstein and Stolle 2008).

According to Flome et al. (2019), when people lose faith in public institutions, the chances of success for disinformation operations increase significantly. In the context of sharing news information online, there is evidence that individuals are more likely to trust, and engage with, a story if it is shared by someone in whom they have a higher level of trust (Sterret et al. 2018). Consequently, the spread of disinformation can be traced back to spreading legitimacy problems in many democracies (Bennett and Livingston 2018). A decline in citizens' confidence in institutions undermines the credibility of official information in the news and makes the public more open to alternative information sources. Many of those sources are often associated with both nationalist and foreign players who aim to undermine institutional legitimacy and destabilise the central parties, governments, and elections. According to Berzins (2018), the main focus for neutralising foreign influence should be on reducing the gap between politics/politicians and the population. One way to achieve this is to carry out analyses to monitor the population's level of openness for information influencing as well as the vulnerabilities that might be exploited. This requires finding out whether society in general, or a specific demographic, shows willingness to defend the country, trust in state institutions, and trust the political and judiciary systems (Berzins 2018).

5 The European Union and Countering Disinformation

Information influencing is not a new phenomenon. However, two occurrences increased the EU's need to develop measures to counter disinformation and propaganda. Firstly, once the so-called Islamic State of Iraq and the Levant (ISIL/Daesh) gained global prominence in early 2014, it became evident that the EU and European citizens were major targets. ISIL engaged in active disinformation campaigns with the aim of undermining European values and interests. These disinformation and propaganda campaigns were spread effectively through traditional and social media, thus enabling the terrorist group not only to create fear, but also to radicalise (European Parliament 2016) and recruit Western citizens, including young Europeans (European Parliament 2015). Secondly, the terrorist attacks in Paris and Brussels as well as the spread of jihadi propaganda signalled a need to re-think the way in which the EU communicates not only with Arab countries, but also with its own citizens within the EU. In both of these cases, the suspects were the citizens of the targeted countries. They had been exposed to jihadi propaganda and were radicalised in Europe. Therefore, designing countermeasures and directing the strategic communications only outside the EU's territory would not be sufficient. The communication channels, language to be used, and key messaging may differ within and outside the EU, which should also be considered when designing the countermeasure toolbox and its implementation.

As a matter of fact, following the start of the Ukraine war, the terms 'hybrid influencing', 'hybrid threat', and 'hybrid warfare' were adopted into EU discourse. Hybrid warfare employs multi-layered efforts designed to destabilise a functioning state and polarise its society (European Commission 2016). The target of hybrid influencing, on the other hand, does not necessarily have to be any clear operation of society. Hybrid influencing may just as well target social communications, trust, and the opinion environment (City of Helsinki 2018). Unlike in conventional warfare, the "centre of gravity" in hybrid warfare is a target population (Caliskan and Cramers 2018, 8). The desired effects may be achieved through the dissemination of falsehoods to undermine morale by creating divisions between the different ethnic, linguistic, and political groupings within these states or by engineering overall dissatisfaction with the government. Perpetrators of hybrid threats can systematically spread disinformation, including through targeted social media campaigns, thereby seeking to radicalise individuals, destabilise society, and control the political narrative (European Commission 2016, 4–5).

In order to build up a resistance against disinformation and propaganda among its member states, in 2015 the EU adopted measures aimed at preventing, identifying, and counteracting hostile actions in the information space. The EU has strengthened its capabilities to identify and counter disinformation, for instance via the East Strategic Communication Task Force, which functions as part of the European External Action Service. The East StratCom Task Force monitors disinformation and increases awareness in Eastern Partnership countries and member states (EEAS 2019a). Another example is the European Centre of Excellence for Countering Hybrid Threats, which was established in Helsinki in 2017. It is not an operational centre for anti-hybrid warfare, but rather a centre that promotes countering hybrid threats at the strategic level through research and training for participants from the EU and NATO. The centre is an excellent example of the deepened cooperation between these two organisations in the field of security and defence (EEAS 2019b). Although the effects of the EU's efforts to counter disinformation and propaganda are somewhat promising, these measures are still not sufficient, as it is challenging to tackle disinformation while protecting fundamental rights, including media pluralism, data protection, and freedom of expression (Syed 2017). In addition, there is still no consensus regarding the best way to counter disinformation at EU level, nor in its member states. Moreover, shared competencies between member states and EU institutions make dealing with the problem even more challenging. Member states have a responsibility to protect their democratic structures, such as elections (Council of the European Union 2019). However, misinformation and disinformation are cross-national issues. Unfortunately, the availability of resources, competences, and existing legal frameworks may differ between individual member states.

The success of disinformation operations depends on their effects on the various target audiences. Therefore, responses should be context-specific and targeted, as well. The 28 EU member states represent a heterogeneous group of countries, whose vulnerabilities differ greatly from one another, making it paramount for the EU members states to develop country-specific countermeasures. Nevertheless, countering disinformation at EU level is also paramount, seeing as national vulnerabilities can have effects that reach beyond borders. Examples for this could be the rise in anti-EU narratives and the spread of falsehoods related to EU membership during the immigration crisis of 2015. An attack against a vulnerable node in one country can have consequences in other countries, where such vulnerabilities previously might not even have existed. A hybrid attack exploiting national vulnerability may therefore require not only sovereign action, but also common planning and a common response (NATO 2018).

Disinformation and misinformation are international issues without geographical borders. Hence, resolving them calls for inter-agency as well as intra-agency action. A mix of different instruments and players need to be employed in order to tackle information influencing effectively. The threat towards national security is increasingly created from within – in versatile, like-minded groups that are being enabled by social media platforms. Consequently, the dissemination of falsehoods seeking to promote overall dissatisfaction with the government or undermine social cohesion by creating divisions between the different ethnic, linguistic, and political groupings within the state should be actively monitored and countered. Employing measures such as strategic communications, transparency, and effectiveness to foster citizens' trust towards public institutions is closely associated with social resilience and general trust (Palmertz 2016). The best defence against information influencing activities is to raise awareness and build societal resilience in the long term.

6 Discussion and Conclusion

Nowadays, social media offers a platform for the dissemination of propaganda, hoaxes, and fake news to dupe the public (Agarwal and Bandeli 2018). It is notable that people frequently continue to rely on misinformation and/or disinformation even after it has been retracted. Social media can also be used as a weapon for destructive and defensive purposes, which are characterised by the term 'information warfare' (Munro 2005). Weaponised information is one way to foster polarisation and consequently to destabilise the population. The best protection against this depends on users' own action and knowledge (Forno 2018). This is particularly important in the case of adolescents who are in the process of constructing their identities and worldviews. Many citizens, particularly young people, use social media applications on a daily basis over long periods of time and may thus be constantly exposed to information influencing. Social media provides massive possibilities in this regard. At the same time, the malicious use of social media is posing more and more of a security threat.

Social media has had a great impact on how armed forces operate (Mangat 2018; Singer and Brooking 2018). Communication platforms on social media have played an increasingly important role for military operations in recent years. Since social media platforms can also be used as a force multiplier, armed forces have changed their attitude towards social media practices. All organisations, including military ones, can use social media platforms to support their day-to-day functions, official communications, networking, and training actions. However, according to Veerasamy and Labuschagne (2018, 47–56), "clear guidance, management and governance are required in order to ensure that the platform is implemented correctly, for example in military organisational processes. Within the military, there may be a vast struggle between the demands for security and the advantages of connectivity and openness."

Indeed, armed forces already have some tools to answer this new social media challenge. The armed forces of the Western world have developed their capacity to monitor open information sources, especially in the media and also increasingly in digital and social media. For example, social media analysis may allow a military operation to target an adversary's movements directly (Peritz 2015). One of the most important tools for military operations is the concept of *Presence, Posture, and Profile* (PPP), which means that military units on all levels have to take care of their actions and movements in the media as well as in the local and global population at all times (NATO 2010). For this reason, it is important to strengthen and support official channels of information as a way of disseminating accurate information.

As individual countries cannot be left to assume sole responsibility in this information war, the EU plays a vital role in supporting Western countries in their fight against disinformation. The EU can support common regulations and also put mechanisms in place, which allow it to steer the overall counteractions necessary to deal with the malicious use of social media.

In a hybrid-threat environment, a society is both an object of aggression and a force that counters aggression. In Finland, for instance, there is a wide perception that a unity-seeking network creates a resilience that can immunise society against hybrid influencing (Aaltola and Juntunen 2018). Generalised trust plays a predominant role in stable societies. That is to say, trust in policy-implementing bodies, such as the police and the armed forces, has a major impact on how individuals perceive their personal and national security. Several conclusions can thus be drawn from the relation between generalised trust and resistance towards disinformation. Firstly, effective resilience requires an open dialogue between public institutions, policymakers, and the population about unfolding hybrid events. Effective communication can help maintaining citizens' trust in liberal values, democratic processes, and governance structures. Secondly, the whole of government – and beyond – is needed to effectively build up resistance. Preparing for hybrid threats, such as misinformation and disinformation, cannot be left to traditional security players alone. Thirdly, there is a need for strategic communication and the clear distribution of tasks. For all the limits regarding what governments can – and should – do, it is important to respond quickly to particular information operations once discovered, both to minimise their impact and to deter other states or groups that might want to emulate the attack.

Furthermore, to better understand the connection between social media and security, it is of paramount importance to understand the plethora of digital social realities by which young people are surrounded, because even adolescents are targets of political, commercial, and social information intended to influence their opinions and actions. Moreover, social media is also exploited by cyberbullies as well as perpetrators of more extreme violence (Norri-Sederholm et al. 2018; Peterson and Densley 2017). As young people live in social media practically twenty-four-seven, it is essential to understand how this has more and more of an effect on national and international security. Thus, it is important to offer general tools to improve media literacy skills as well as increase the understanding of information influencing and the dark side of social media. What is more, increasing and maintaining trust in society in general and in national security institutions, such as the police and the armed forces, in particular is of the utmost importance. The national security institutions in Finland, for instance, have made efforts to generate trust in these institutions. The authorities use social media and citizens can, whenever they have a need, chat with the police. Moreover, the Finnish Defence Forces have developed a special social-media system through which conscripts can discuss with other soldiers all kinds of problems they might have during their service time.

It is difficult to single out individual contributors or culprits when it comes to the dissemination of misinformation and disinformation, due to the ambivalent nature of the internet and the Western approach to using social media. What is more, principles such as ease of access to the internet, democracy, freedom of speech, and knowledge equality also make it easier to distribute disinformation.

This contribution aimed to understand the flow of misinformation and disinformation in social media in relation to armed forces and national security. In this respect, we suggest that it is possible to create generalised trust towards public authorities both nationally and internationally, for example by means of EU initiatives. Furthermore, by describing aspects of misinformation and disinformation in social media from a national security point of view, we were able to show that international cooperation as well as the role of the EU are important for armed forces on this particular battlefield. Understanding the different dimensions and challenges information warfare poses to armed forces both in peacetime and wartime conditions is crucial. Finally, when thinking of national security, it is important to remember the young people – our future conscripts and soldiers – and to strengthen their abilities to survive in a world of misinformation and disinformation.

Acknowledgments This study is part of a research project funded by the Academy of Finland.

References

Aaltola M, Juntunen T (2018) Nordic model meets resilience – Finnish strategy for societal security. In: Sprūds A, Kuznetsov B, Aaltola M (eds) Societal security in the Baltic Sea region. Latvian Institute for Foreign Affairs, Rīga, pp 26–42

Abramson J (2017) Trust and democracy, knight commission on trust, media, and American democracy. The Aspen Institute, July 2017 [online]. [Viewed 4 February 2020]. Available from: https://www.aspeninstitute.org/programs/communications-and-society-program/knight-commission-workshop-trust-media-american-democracy-2/

Agarwal N, Bandeli KK (2018) Examining strategic integration of social media platforms in disinformation campaign coordination. Def Strateg Commun 4(Spring):173–206

Bäck M, Kestilä-Kekkonen E (eds) (2019) Political and social trust: pathways, trends and gaps (in Finnish). Publications of the Ministry of Finance 2019, (31)

Bakir V, McStay A (2018) Fake news and the economy of emotions: problems, causes, solutions. Digit Journal 6(2):154–175

Bennett L, Livingston S (2018) The disinformation order: disruptive communication and the decline of democratic institutions. Eur J Commun 33(2):122–139

Berg M, Johansson T (2016) Trust and safety in the segregated city: contextualizing the relationship between institutional trust, crime-related insecurity and generalized trust. Scand Polit Stud 39(4):458–448

Berzins J (2018) Public hearing of the Select Committee on deliberate online falsehoods. Channelnewsasia [online]. [Viewed 27 June 2019]. Available from: https://www.channelnewsasia.com/news/singapore/society-s-mental-state-a-key-battleground-for-misinformation-10045296

Blind PK (2006) Building trust in government in the twenty-first century: review of literature and emerging issues. UNDESA 7th Global Forum on Reinventing Government, 26–29 June 2007, Vienna. United Nations Department of Economic Social Affairs (UNDESA), New York

Boda Z, Medve-Balint G (2017) How perception and personal contact matter: the individual determinants of trust in police in Hungary. Polic Soc 27(7):732–749

Caliskan M, Cramers P-A (2018) What do you mean by "Hybrid Warfare"? A content analysis on the media coverage of hybrid warfare concept. Horiz Insights 1:23–35

Chen X, Sin SCJ, Theng YL, Lee CS (2015) Why students share misinformation on social media: motivation, gender, and study-level differences. J Acad Librariansh 41(5):583–592

City of Helsinki (2018) Helsinki in the era of hybrid threats – hybrid influencing and the city, vol 22. Publications of the Central Administration, Helsinki

Committee on Foreign Relations (2018) Putin's asymmetric assault on democracy in Russia and Europe: implications for U.S. national security. United States Senate, Washington, DC

Council of the European Union (2019) Complementary efforts to enhance resilience and counter hybrid threats. Council Conclusions 10 December 2019

Del Vicario M, Bessi A, Zollo F, Petroni F, Scala A, Caldarelli G, Stanley HE, Quattrociocchi W (2016) The spreading of misinformation online. Proc Natl Acad Sci 113(3):554–559

DiResta R, Shaffer K, Ruppel B, Sullivan D, Matney R, Fox R, Albright J, Johnson B (2018) The tactics & tropes of the internet research agency. New Knowledge, New York. [online]. Available from: https://disinformationreport.blob.core.windows.net/disinformation-report/NewKnowledge-Disinformation-Report-Whitepaper-121718.pdf

Dubois E, Blank G (2018) The echo chamber is overstated: the moderating effect of political interest and diverse media. Inf Commun Soc 21(5):729–745

EEAS (2019a) Countering disinformation, European External Action Services [online]. [Viewed 28 June 2019]. Available from: https://eeas.europa.eu/topics/countering-disinformation/59411/countering-disinformation_en

EEAS (2019b) EU-NATO cooperation – factsheets, European External Action Services [online]. [Viewed 28 June 2019]. Available from: https://eeas.europa.eu/headquarters/headquarters-Homepage/28286/eu-nato-cooperation-factsheet_en

European Commission (2016) Joint framework on countering hybrid threats – a European Union response. Joint Communication to the European Parliament and the Council, Brussels, 6 April 2016

European Parliament (2015) Understanding propaganda and disinformation. At a glance November 2015. European Parliament [online]. [Viewed 15 May 2019]. Available from: http://www.europarl.europa.eu/RegData/etudes/ATAG/2015/571332/EPRS_ATA(2015)571332_EN.pdf

European Parliament (2016) EU strategic communication with the Arab world. Briefing May 2016. European Parliamentary Research Service [online]. Available from: http://www.europarl.europa.eu/RegData/etudes/BRIE/2016/581997/EPRS_BRI(2016)581997_EN.pdf

Fallis D (2014) The varieties of disinformation. In: Floridi L, Illari P (eds) The philosophy of information quality. Springer, Cham, pp 135–161. https://doi.org/10.1007/978-3-319-07121-3_8

Fallis D (2015) What is disinformation? Libr Trends 63(3):401–426. https://doi.org/10.1353/lib.2015.0014

Fetzer JH (2003) Information: does it have to be true? Mind Mach 14(2):223–229

Fetzer JH (2004) Disinformation: the use of false information. Mind Mach 14(2):231–240

Flome M, Blahur A, Podavini A, Verile M (2019) Understanding citizens' vulnerabilities to disinformation and data-driven propaganda: case study: the 2018 Italian general election, JRC technical report, European Commission, JRC116009: EUR 29741 EN. Publications Office of the European Union, Luxembourg

Floridi L (2011) The philosophy of information. Oxford University Press, Oxford

Forno R (2018) Weaponized information seeks a new target in cyberspace: users' minds. The Conversation. [Online]. [Viewed 15 May 2019]. Available from: https://theconversation-com.cdn.ampproject.org/c/s/theconversation.com/amp/weaponized-information-seeks-a-new-target-in-cyberspace-users-minds-100069

Fukuyama F (2001) Social capital, civil society and development. Third World Q 22(1):7–20

Funke D, Flamini D (2020) A guide to anti-misinformation actions around the world. Poynter. [Online]. [Viewed 10 January 2020]. Available from: https://www.poynter.org/ifcn/anti-misinformation-actions/

Giacometti A, Teräs J, Perjo L, Wøien M, Sigurjonsdottir H, Rinne T (2018) Regional economic and social resilience: conceptual debate and implications for Nordic regions. Discussion paper prepared for Nordic thematic group for innovative and resilient regions, January 2018, Stockholm

Gripsrud J, Moe H, Molander A, Murdock G (eds) (2010) The idea of the public sphere: a reader. Lexington Books, Plymouth

Guess A, Nyhan B, Reifler J (2018) Selective exposure to misinformation: evidence from the consumption of fake news during the 2016 US presidential campaign. European Research Council. [Online]. [Viewed 18 May 2019]. Available from: http://www.ask-force.org/web/Fundamentalists/Guess-Selective-Exposure-to-Misinformation-Evidence-Presidential-Campaign-2018.pdf

Gutierrez R (2018) Tackling disinformation in the age of social media. Available via Equal times. [Online]. [Viewed 10 January 2020. Available from: https://www.equaltimes.org/tackling-disinformation-in-the-age#.Xhqv28gzZPY

Håkansson P, Witmer H (2015) Social media and trust — a systematic literature review. J Bus Econ 6(3):517–524

Hybrid CoE (2019) Countering disinformation: news media and legal resilience. COI Records

Karlova NA, Fisher KE (2013) "Plz RT": a social diffusion model of misinformation and disinformation for understanding human information behaviour. Inf Res 18(1):1–17

Kasher A (2003) Public trust in a military force. J Mil Ethics 2(1):20–45

Kim KS, Sin SCJ, Yoo-Lee EY (2014) Undergraduates' use of social media as information sources. Coll Res Libr 75(4):442–457

Kofman M, Migacheva K, Nichiporuk B, Radin A, Oberholtzer J (2017) Lessons from Russia's operations in Crimea and eastern Ukraine. RAND Corporation, Santa Monica

Kumar KK, Geethakumari G (2014) Detecting misinformation in online social networks using cognitive psychology. Human-centric Comput Inf Sci 4(1):14

Levi M (1998) A state of trust. In: Braithwaite V, Levi M (eds) Trust & governance. Russell Sage Foundation, New York

Levi M, Stoker L (2000) Political trust and trustworthiness. Annu Rev Polit Sci 3:475–507

Limnéll J, Rantapelkonen J (2017) Pelottaako? Nuoret ja turvallisuuden tulevaisuus (Are you afraid? Young people and the future of safety and security). Docendo, Helsinki

Lotti M (2018) Countering misinformation online: policies and solutions. Report from 14 November 2018. Geneva Internet Platform Digital Watch Observatory. [Online]. [Viewed 10 January 2020]. Available from: https://dig.watch/sessions/countering-misinformation-online-policies-and-solutions

Lyngaas S (2018) Russian disinformation ops were bigger than we thought. Cyberscoop. [Online]. [Viewed 15 May 2019]. Available from: https://www.cyberscoop.com/russian-information-operations-senate-intelligence-committee/

Mangat R (2018) Tweeting strategy: military social media use as strategic communication. Theses and Dissertations (Comprehensive), Wilfrid Laurier University. 2071. [Online]. [Viewed 13 May 2019]. Available from: https://scholars.wlu.ca/etd/2071

Munro I (2005) Information warfare in business. Strategies of control and resistance in the network society. Routledge, London

Nannestad P (2008) What have we learned about generalized trust, if anything? Annu Rev Polit Sci 11:413–436

NATO (2010) NATO Bi-SC information operations reference book. Version 1. 05 March 2010 NATO/PfP/Unclassified Releasable to EU/NATO Partner Nations/ISAF/KFOR/OAE/OOS [online]. Available from: https://info.publicintelligence.net/NATO-IO-Reference.pdf

NATO (2018) Cooperating to counter hybrid threats. NATO review published on 23 November 2018 [online]. [Viewed 8 May 2019]. Available from: https://www.nato.int/docu/review/2018/Also-in-2018/cooperating-to-counter-hybrid-threats/EN/index.htm
Newton K (2001) Trust, social capital, civil society, and democracy. Int Polit Sci Rev 22(2):201–214
Newton K (2007) Social and political trust. In: Dalton RJ, Klingemann HD (eds) The Oxford handbook of political behaviour. Oxford University Press, Oxford
Norri-Sederholm T, Huhtinen A-M, Paakkonen H (2018) Ensuring public safety organisations' information flow and situation picture in hybrid environments. Int J Cyber Warf Terror 8(1):12–24
Offe C (2000) Trust, democracy and justice. Theoria J Soc Polit Theo 96(December):1–13
Palmertz B (2016) Europeiska Perspektiv På Förmågan Att Möta Påverkanskampanjer Från Främmande Makt – Delrapport 1. (European perspective on the ability to counter foreign power influence campaigns). Center for Asymmetric Threat Studies (CATS), Swedish National Defence College, Stockholm
Pamment J, Agardh-Twetman H (2018) The role of communicators in countering the malicious use of social media. NATO Strategic Communications Centre of Excellence
Pamment J, Nothhaft H, Agardh-Twetman H, Fjällhed A (2018) Countering information influence activities: the state of the art, version 1.4 (1 July 2018). Department of Strategic Communication, Lund University, Lund
Peritz A (2015) Yes, the U.S. Military uses social media to target ISIS. Slate, June 11 2015. [Online]. [Viewed 12 May 2019]. Available from: https://slate.com/news-and-politics/2015/06/the-u-s-military-uses-social-media-to-target-isis-stop-reminding-the-jihadis.html
Peterson J, Densley J (2017) Cyber violence: what do we know and where do we go from here? Aggress Violent Behav 3:193–200
Putnam, RD (1993) The Prosperous Community. The American Prospect 4(13):35–42
Ratkiewicz J, Conover M, Meiss M, Gonçalves B, Patil S, Flammini A, Menczer F (2010) Detecting and tracking the spread of astroturf memes in microblog streams. arXiv preprint arXiv, 1011.3768
Rothstein B, Stolle D (2003) Social capital, impartiality, and the welfare state: an institutional approach. In: Hooghe M, Stolle D (eds) Generating social capital: the role of voluntary associations. Institutions and government policy. Palgrave Macmillan, New York
Rothstein B, Stolle D (2008) The state and social capital. Comp Polit 40(4):441–459
Sanovich S (2017) Computational propaganda in Russia: the origins of digital misinformation. Computational Propaganda Research Project. Working paper (3)
Security Strategy for Society (2017) Resolution of the Finnish government Vol 2. The Security Committee. [Online]. [Viewed 19 January 2019]. Availiable from: https://turvallisuuskomitea.fi/wp-content/uploads/2018/04/YTS_2017_english.pdf
Singer PW, Brooking ET (2018) LikeWar. The weaponization of social media. An Eamon Dolan book. Houghton Mifflin Harcourt, New York
Smith NR (2015) The EU and Russia's conflicting regime preferences in Ukraine: assessing regime promotion strategies in the scope of the Ukraine crisis. Eur Secur 24(4):525–540
Søe SO (2018) Algorithmic detection of misinformation and disinformation: Gricean perspectives. J Doc 74(2):309–332
Sterret D, Malato D, Benz J, Kantor L, Tompson T, Rosenstiel T, Sonderman J, Loker K, Swanson E (2018) Who shared it? How Americans decide what news to trust on social media. NORC Working Paper Series: WP-2018-001
Syed N (2017) Real talk about fake news: towards a better theory for platform governance. Yale Law J 127(Forum):337–357
Tarrow S (1996) Making social science work across space and time: a critical reflection on Robert Putnam's making democracy work. Am Polit Sci Rev 90(3)
The Law Library of Congress (2019) Initiatives to counter fake news in selected countries. Law Library – Library of Congress. [Online]. [Viewed 10 January 2020]. Available from: https://www.loc.gov/law/help/fake-news/counter-fake-news.pdf

Thompson I, Lapowsky I (2018) How Russian trolls used meme warfare to divide America. Wired Security [online]. [Viewed 11 May 2019]. Available from: https://www.wired.com/story/russia-ira-propaganda-senate-report/

Uslaner EM (2002) The moral foundation of trust. Cambridge University Press, New York

Uslaner EM, Brown M (2005) Inequality, trust, and civic engagement [online]. Am Polit J 33(6):868–894. Available from: https://doi.org/10.1177/1532673X04271903

Veerasamy N, Labuschagne WA (2018) Framework for military applications of social media. Int J Cyber Warf Terror (IJCWT) 8(2):47–56. https://doi.org/10.4018/IJCWT.2018040104

WestDM(2017)Howtocombatfakenewsanddisinformation[online].[Viewed11May2019].Available from: https://www.brookings.edu/research/how-to-combat-fake-news-and-disinformation/

Woolley SC, Howard P (2017) Computational propaganda worldwide: executive summary. Working paper 11. Oxford, UK, Project on Computational Propaganda

World Economic Forum (2014) The rapid spread of misinformation online [online]. [Viewed 8 March 2017]. Available from: http://wef.ch/GJAfq6

Zak PJ, Knack S (2001) Trust and growth. Econ J 111(470):295–321

Social Media Use in Contemporary Armed Forces as a Mixed Blessing

Jelena Juvan and Uroš Svete

Abstract The widespread use of social media in the armed forces has caused structural changes in the armed forces' communications with both the wider public and the internal public. While social media has undoubtedly eased communication with the home front, several hidden dangers can be identified. The following paper reflects on the dangers and benefits of the use of social media platforms for the armed forces from two different perspectives, the institutional and the individual. This paper also tries to answer the question of how we can regulate something that cannot be regulated, and identifies positive and negative consequences of social media use in and by the armed forces.

Keywords Social media · Armed forces · Regulations · Slovenian Armed Forces · Information sharing

1 Introduction

The dramatic rise in social media use has become one of the main characteristics in contemporary societies in the last decade. Social media has become an important communication tool not only for personal use, but also for institutional purposes. Armed forces, which as traditional organisations are by nature slow in adapting to changes and accepting novelties, have been rather hesitant in adopting social media as a communication tool. One of the issues connected with the use of social media in the armed forces is the danger of over-sharing. This does not only refer to the over-sharing of information, but also to sharing general information – intentionally or unintentionally – which is meant to be kept secret. In November 2017, U.S. fitness-tracking company Strava released a global heatmap "as a visualisation of two years

J. Juvan (✉) · U. Svete
Faculty of Social Sciences, University of Ljubljana, Ljubljana, Slovenia
e-mail: jelena.juvan@fdv.uni-lj.si

E. Moehlecke de Baseggio et al. (eds.), *Social Media and the Armed Forces*, Advanced Sciences and Technologies for Security Applications,
https://doi.org/10.1007/978-3-030-47511-6_13

of trailing data from Strava's global network of athletes" (Robb 2017). Unintentionally, sensitive information about the locations and staffing of military bases and spy outposts around the world was revealed:

> The map shows every single activity ever uploaded to Strava – more than 3 trillion individual GPS data points, according to the company. The application can be used on various devices including smartphones and fitness trackers to see popular running routes in major cities, or spot individuals in more remote areas who have unusual exercise patterns (Hern 2018).

However, the published map gave away much more than the company intended. It was detailed enough to potentially disclose extremely sensitive information about a subset of Strava users: military personnel on active service. Analysts noted that in locations such as Afghanistan, Djibouti, and Syria, the users of Strava seem to be almost exclusively foreign military personnel, meaning that bases stand out brightly on the map.

The above-mentioned case is not directly connected with the use of social media. However, it is a good example of the type of risks new information technologies may pose when various personal data and info is published online. Moreover, these risks are even higher when military organisations and military personnel are involved. A soldier who goes for the daily run inside the military base somewhere in the middle of Afghanistan and turns on the Strava application to keep track of his or her physical activity surely does not think about the possible consequences of GPS tracking the route, let alone that the tracking app might be giving away top secret information. With the widespread use of social media platforms, opportunities to share information are unlimited, which also entails several dangers and risks. For instance, it is not just the opportunities that are unlimited, but also the reasons for publishing and sharing information online. "Every person has a button. For somebody there's a financial issue, for somebody it's a very appealing date, for somebody it's a family thing" (Lapowsky 2019). When trying to control social media, we are actually trying to control individuals. The question is, how can we regulate what is being published online by military personnel when we cannot regulate social media itself? The use of social media can be monitored and partially restricted in terms of confidentiality of information and security of the execution of tasks in peace and war times.[1] For this reason, the debate on social media use in the armed forces and by the armed forces is necessary.

With regard to social media use and the armed forces, two perspectives can be identified: firstly, the institutional perspective, meaning the use of social media as a formal communication channel with the public; and secondly, the personal perspective, which refers to the use of social media by individual members of the armed forces.

---

[1] It is not only about the information we share willingly, it is also about the privacy features and settings of social media platforms, which, according to researchers, cannot be trusted not to leak information to other layers of the social media platform or to other users and companies with an interest in such information (Bay et al. 2019).

2 Methodology

The present paper is divided into two parts. The first part shows an overview of the process of recognising the importance of social media for the armed forces, including some theoretical concepts which can be applied when analysing the use of social media as a new phenomenon.

The second part focuses on an empirical study by the Slovenian Armed Forces (SAF) and the Slovenian Ministry of Defence (MoD). For this study, structured interviews were conducted with chosen personnel in the SAF and MoD. The interviews were planned and conducted in order to follow the two aforementioned perspectives. The institutional perspective was covered by means of a group interview with personnel in the Department for Recruitment in the Military Affairs Division (Defence Affairs Directorate) at the MoD of the Republic of Slovenia. These personnel are responsible for managing the social media handle *Postani vojak* (*'become a soldier'*), whose sole purpose is to promote the benefits of employment in the SAF for the purposes of recruitment (Interview 1 2019).

For the individual perspective, interviews were conducted with two members of the SAF. The first interviewee was major Nina Raduha[2], an experienced member of the SAF, who was, among other things, also the first female commander of a contingent in the UNIFIL mission (Interview 2 2019). Her experiences also include the position of public relations officer and human intelligence officer, which makes her a very suitable person to interview for the purpose of this article. The second interview was conducted with a member of the SAF in the position of Chief of Staff, who expressed the wish to stay anonymous (Interview 3 2019). He is an experienced service member who has been employed in the SAF since 2002 and has been deployed to missions abroad several times, including a mission to Afghanistan. He was chosen due to his strong presence on social media as well as due to the fact that he was a commanding officer during his deployments and is thus very familiar with soldiers under his command over-sharing information on social media. For both interviews, an official request was filed and later approved by the responsible commander.

[2] Major Nina Raduha, B.A in Political Science – Defence studies. She holds a master's degree in criminal law from the Faculty of Law, University of Ljubljana. She has been employed in the SAF since 2003 and is an infantry officer. She participated in missions in Bosnia and Herzegovina and Lebanon. In 2016, she completed her senior staff training at the top of her class.

3 Traditional Armed Forces and "Non-traditional" Social Media

The phenomenon of social media has thoroughly changed the way people and institutions communicate, not only among each other, but also with the public. "In the present digital world, social media is the new frontier for signal intelligence" (Gupta 2017). The armed forces of the United States, Israel, Australia, France, Germany, and many other Western countries maintain an online presence across various digital social media platforms, as do the armed forces of non-Western states such as Pakistan, India, Russia, and China. One could say that all of the world's main military actors have acknowledged the use and the importance of social media platforms. As Gupta (2017) notes, contemporary armed forces "cannot be isolated from usage of these platforms in today's digital world. With its unprecedented reach as well as constructive and destructive powers, it is imperative that the armed forces learn to harness this potent tool for achieving organisational goals."

Social media can be also considered as a form of soft power. The concept of soft power was introduced by Joseph S. Nye in 1990. He describes soft power as "getting others to want the outcomes that you want. [...] A country may obtain the outcomes it wants [...] because other countries – admiring its values, emulating its example, aspiring to its level of prosperity and openness – want to follow it" (Nye 2004, 5). Furthermore, soft power tends to be "associated with intangible power resources such as culture, ideology and institutions" (Nye 1990, 166). The media in general is considered a soft power. Thus, new types of media such as social media can also be considered as soft powers, seeing as they are used to influence people's way of thinking. The case of Cambridge Analytica in 2014 is proof of the power of social media.

Having recognised the soft power and the influence of this new medium, the major military powers in the world have undergone an evolution process while laying down guidelines for the use of social media. A 2010 U.S. Department of Defence (DoD) memo states that "Internet based capabilities are integral to operations" (Gupta 2017). As such, the U.S. Armed Forces realise that social media enables them to communicate in new ways to a larger audience and faster than ever before – and, most importantly, without any distortion of the content (Gupta 2017).

As the use of social media increases exponentially[3], armed forces authorities have to devise means to exploit these platforms to their advantage in numerous fields. The purpose as to why armed forces use social media platforms and communication applications is two-tiered. Firstly, social media is used to communicate with the public, informing them about the tasks, duties, and everyday happenings inside the armed forces, which is important for obtaining social legitimacy. Secondly, it is used for official, internal communication among members of the armed forces.

[3]The number of internet users worldwide in 2019 was 4388 billion, which is up 9.1% year-on-year. The number of social media users worldwide in 2019 was 3484 billion, up 9% year-on-year (Chaffey 2019).

For the purposes of official, internal communication, Facebook (Messenger) is not appropriate, since it is not considered safe. Communication via Viber, Telegram, Signal, Skype or WhatsApp[4] is safer, since these services offer end-to-end encryption. Moreover, Viber, Signal, and Telegram offer the possibility of secret chats and setting a timer that deletes the message at the specified time. Therefore, several armed forces have been using these applications as official communication channels. For example, as was stated by one of the interviewees who cooperated with the Canadian Armed Forces while being deployed, Canadian forces on missions abroad used WhatsApp to share official orders, security alerts, information, and so forth (Interview 3 2019). A completely opposite stance regarding the use of social media is adopted by the Russian Armed Forces, who decided to ban the use of smartphones for soldiers while they are on duty, based on the issues of national security raised due to their use of social media platforms (BBC News 2019).[5] The same approach can be identified in the Indian armed forces (Gadgets Now 2020).

Using social media is not just about communication and spreading the news. It is also an opportunity for military professionals to extend their span of influence beyond the chain of command, cut through multiple layers of bureaucracy, and potentially develop a personal form of soft power. Social media offers a vast array of possibilities and it is down to a user's own capabilities how well he or she will be able to use it and for which purposes.

For decades, social institutions such as the military have been described as *greedy* institutions (Coser 1974), placing extensive demands and high expectations on their members as well as impacting their lives in multiple domains such as work, family, and social relationships outside of duty hours (Hatch et al. 2013). As an institution, the military is structured for its members to develop a strong identity, which is derived from a degree of internal group cohesion, in order to sustain a force capable of engaging in war – the ultimate purpose of any armed force – as well as the individual motivation to fight. This is achieved by installing distinct cultural traditions and rituals, affirming the primacy of the group and the institution over their members through an open-ended liability for service, and in return providing them with institutional support and care (Moskos 1986). This level of institutionalism is further enhanced by what Coser (1974) identified as an important characteristic of greedy institutions, namely that they rely on voluntary compliance and, at the same time, exert pressure on members to break ties with other institutions or individuals who might make conflicting demands.

With their specific professional culture, armed forces are considered as some of the most traditional organisations. Wiatr (1987, 34–35) defines the basic characteristics of the armed forces as an institution as follows:

[4]WhatsApp is owned by Facebook. The merging of FB Messenger, Instagram, and WhatsApp is planned for the near future.

[5]In recent years, social media posts by servicemen have revealed Russia's military presence in eastern Ukraine and Syria, sometimes contradicting the government's official claim of not having troops stationed there.

> Armed forces are organisations in which professional ties prevail over personal ties. They are bureaucratic organisations with a hierarchical structure. Armed forces are a social environment with its own system of stratification; a battle group organised around the purpose of winning the battle.[6]

The characteristics of the armed forces influence and shape the main characteristics of the employed personnel. With regards to life in the armed forces, Segal (1986, 16–22) identifies four main characteristics: "risk of injury or death", "geographic mobility", "separations", and "residence in foreign countries". Specific characteristics of the armed forces as an organisation strongly influence all areas, including communication with the public. Traditionally, armed forces were a closed social group whose life was limited to military barracks and military bases, without frequent interactions with the civilian world or sharing information to the public. Now, due to the development of the Information Communication Technology (ICT) and the wide spread of social media, this is no longer the case. Whether they like it or not, the armed forces have to talk to the 'outside' world.

Hajjar (2014) emphasises the impact of the information age on the culture of the U.S. Armed Forces. The information age has helped change how the military views the world. However, while the rise of the concept of network-centric warfare in military doctrine has strong technological connotations, it also betrays social preoccupations, such as the importance of understanding, and effectively influences a variety of people's networks. In order to deal with these challenges, the U.S. military hired thousands of civilians who are experts in the use of new sophisticated systems, and active-duty military members are being taught new information skills. Thus, for these reasons and others, the information age has substantially changed the organisation and has contributed to an emergent post-modern military culture (Hajjar 2014).

The debate over the extent to which the military should incorporate rapidly expanding and changing ICT highlights another disputed aspect of military culture, as it also applies to social media issues. Over the past decade, the use of social media by U.S. Armed Forces personnel went from being banned to being openly embraced, used, and promoted by DoD officials. In 2007, the Pentagon blocked its computer networks from accessing sites, including YouTube and MySpace, in an attempt to keep a tight rein on information about troop activities. Official documents at the time cited not only threats to safety but also the heavy burden placed on bandwidth. In August 2009, the U.S. Marine Corps issued a ban on the use of social networking sites, expressing concerns about the possibility of security risks. Marines were informed that they could no longer use sites such as Facebook, MySpace, and Twitter due to the possibility that enemy groups might use to their advantage the information they contain (Chalmers 2011; Paganini 2013). Although these bans were eventually lifted, the Pentagon established and allowed the launch of

[6] This quote was translated into English by the authors of this paper.

TroopTube[7], a military-sponsored version of YouTube, which enabled the DoD to monitor the flow of information and allowed serving personnel to share videos with their friends and family members only.

In 2010, U.S. Deputy Defence Secretary William Lynn ordered a review of the military's social networking policies to compare the benefits and drawbacks of allowing members of the armed forces to use social media. Later that year, the Pentagon announced it had re-authorised the use of Twitter, Facebook, and other Web 2.0 sites across the U.S. Armed Forces, arguing that the benefits of social media outweighed security concerns. Since 2010, the Pentagon has actively supported the use of social media in many of its efforts. While restrictions still exist, they do not significantly differ from those in place at any other major private company (Matthews-Juarez et al. 2013).

The use of ICT for personal communication first appeared during the American deployment in Somalia in 1993, when U.S. soldiers wished to be able to use communication media, but none was readily available (Ender 2009). Limited telephone use overcame the lack of telecommunication services in and around Somalia and alleviated the stress for soldiers and their families. Problems with communication between the front line and home precipitated an innovative email programme that pleased many spouses by ultimately providing the speed, relative privacy, decentralisation, and personal communication the soldiers desired (Ender 2009). More than two decades after the deployment in Somalia, the use of ICT became a widespread phenomenon among American soldiers during the war in Iraq (Ender 2009). In 2011, the U.S. Army published its *Social Media Handbook* and posted it online.

For one thing, the U.S. Armed Forces recognised that social media has become an important tool for messaging and outreach. Soldiers have always been the military's most effective messengers. Every time a soldier joins the armed forces' social media, it increases the military's capacity for the timely and transparent dissemination of information. Moreover, it ensures that the military's story is shared directly with American citizens wherever they are and whenever they might want to read or hear it. In other words, social media allows the U.S. Armed Forces to connect with its members, and the civilian U.S. with its military.

[7] TroopTube was an online video-sharing platform of the U.S. Armed Forces via their organisation Military OneSource, established to help military families connect and keep in touch while miles apart. TroopTube allowed people to register as members of one of the branches of the armed forces, family, civilian DoD employees or supporters. Members could upload personal videos from anywhere with an Internet connection. The videos were reviewed for content based on the terms of service before being posted to the site. In May 2009, TroopTube was recognised by the White House office of New Media as an innovative way for people to connect with troops, and was featured in a video produced by the White House. While access to TroopTube was blocked at some military bases, it was accessible via many other military bases and available to military families around the world. On 31 July 2011, The Office of the Secretary of Defense closed the TroopTube website, citing the decline in use as a result of the DoD's social media policy implemented in February 2010, which permitted wider access to social media and video-sharing platforms (Shachtman 2009).

For another thing, and even more importantly, it is a fairly well-documented fact that communication between the field and home is most highly valued by soldiers while deployed overseas (Applewhite and Segal 1990; Ender 2009). More frequent and direct contact with the home-front can boost soldiers' morale, help solve family issues before they get out of control, minimise boredom, and effectively ease families' worries regarding the security and wellbeing of their deployed family members (Bell et al. 1999). Offering appropriate and sufficient information to families at home adds to the quality of soldiers' relationships both with their units and their commanding officers (Bartone 2005). However, this is a double-edged sword, as too much information can have the opposite effect. There is always the risk of family receiving bad news through social media instead of official institutional communication channels.[8] Furthermore, worried spouses waiting at home can be bombarded with all different types of information, which can influence the everyday life of the whole family (ibid.). Thus, the same social media that allows families to share intimate moments and events can also heighten emotions and alter behaviour when it is used to inform a soldier of a death in the family, discuss paying bills, address child-rearing problems or mental health issues, including substance abuse and domestic violence. It is also through social media that depression, longing, loneliness, and other feelings of deprivation are voiced and shared (Dao 2011). Nevertheless, the use of social media can generally minimise boredom among deployed soldiers, which has long been a central feature of the military due to under-utilisation, cultural deprivation, lack of privacy, and isolation in time and space (Ender 2009).

While in recent years social scientists have paid increasingly more attention to how social networks influence military family life, the dilemma as to whether the use of ICT and digital social media should be allowed in the armed forces has not been entirely resolved (Matthews-Juarez et al. 2013). The military's use of social media has been widely discussed, often based on those aspects of military culture that militate against its use. Jones and Baines (2013) explored how military commanders attempt to control it, taking into consideration how social outcomes might be pursued if the military conceded to slight conceptual and organisational adjustments. Typically, this culminated in recommendations for the delegation of blogging to the lowest possible level, trading risk for relevance and responsiveness. Lists of factors to be considered in making communication via social media a success have emerged across the Internet along with tips for 'new media literacy' and authoring.

In some cases, there is a feeling that armed forces do not wish to use social media; that they do not have a clear picture of what to do on social media platforms or what type of information they wish to share with the public. Compared to the

[8] Svete and Juvan (2016) report on the case of the SAF soldier serving in Afghanistan who was shot in the leg. The incident was reported as having occurred during the performance of regular duties, thus insinuating that the soldier had shot himself by accident. This put the SAF Special Forces' military skills and professionalism into question. Soldiers in Afghanistan were unable to understand how such imprecise information was fed to the Slovenian public. Reading domestic comments on forums and other social media, some soldiers felt sad, disappointed, and angry.

leading armed forces of the world who have fully grasped the potential of new communication technologies, we can also identify cases (for example, the Slovenian Armed Forces) where the unlimited range of modern communication technologies and all the facts that follow from the changed security environment and the emerging challenges are not fully understood yet.

4 The Slovenian Armed Forces and the Use of Social Media: Institutional and Individual Perspectives

Despite the fact that the SAF have been present on social networks since 2012, which can be considered rather early for this type of organisation, this area is still being left to the work and energy of only one or two individuals (Interview 2 2019), and has not been systematically addressed by the organisation (Interview 1 2019).

The SAF present themselves on social media platforms with two different profiles. The first one is *Slovenska vojska*[9], which is mainly intended for informing the general public about the activities of the armed forces. The second one is *Postani vojak*[10], which is intended solely for the recruitment of new personnel. The activities of the two profiles are strictly separated, which is also reflected in the organisational perspective. The social media accounts of *Slovenska vojska* are administered by personnel in the Public Affairs Department of General Staff, while those of *Postani vojak* are administered by staff members of the Personnel Recruitment Department in the Military Affairs Division at the MoD of the Republic of Slovenia. This rather new division was formed in 2007 based on the experiences of some foreign armed forces who also separated recruitment efforts from public relations (for example, Go Army in the U.S. Armed Forces). While the retention of employed personnel has also become a major issue in the Slovenian Armed Forces during the past years, it should be noted that *Postani vojak* is meant for recruitment purposes only. Thus, all its activities on social media are intended solely for the recruitment of new personnel.

Postani vojak is currently active on four social media platforms: YouTube, Facebook, Instagram, and LinkedIn. The decision to be present on each one of these platforms was not based on a strategic communication reflection, but mostly due to individual decisions as well as popular trends among the population aged 18–27, which is the target demographic for recruitment. Due to financial and human resources shortages, the interviewees for this paper estimate that no additional social media platform could be administered even if it did become very popular among the targeted population (for example Snapchat). However, the SAF do follow the trends and are familiar with new platforms and forms of social media (Interview 1 2019). The main problem with social media use, according to the interviewees, is the product they are 'trying to sell', namely employment in the SAF,

[9] http://www.slovenskavojska.si.

[10] https://www.postanivojak.si.

which is currently being considered a 'bad product' in Slovenian society in general and among the targeted population in particular. Based on these statements, one of the major conclusions of this small-scale research is the identification of inconsistencies in the communication strategy between the main institutional public relations branches. This draws attention to the problems in the armed forces which in turn weakens the brand and operational or tactical public relations wishing to promote a more positive picture of the SAF as well as gaining new staff.

Another institutional issue identified in the interviews is the absence of single units of the SAF on social media (Interview 3 2019). Single units and certain parts of the SAF are not allowed to communicate with the public and therefore are not allowed to have their own Facebook, Twitter or other accounts. According to one interviewee, having…

> …single SAF units present on social media platforms, informing the public about their everyday work and tasks, would make the public more informed. […] All in all, this may be one of the solutions for the serious recruitment issues the SAF have been facing during the last years (Interview 3 2019).[11]

With regard to the use of social media platforms by individual SAF members, thus far no regulations exist. The use of these platforms depends upon and is solely up to the individual's personal culture and awareness of their online activities. The main question, then, is whether there even is a necessity for any type of regulations to cover the individual use of social media platforms for service members. Here, two opposite opinions were detected during the course of the interviews. Firstly, there is a view that the personal use of social media is exclusively a private matter, which should not – and must not – be regulated by institutional rules. In fact, any…

> …kinds of rules and regulations for the members of the armed forces are not necessary, since the use of social media for members is limited to their private lives, and they are using social media as individuals, not as service members (Interview 3 2019).

However, this view clashes with one of the main characteristics of the military job, which is its greediness. That is to say, a soldier is a soldier even when he or she is off duty. Is it thus even possible to distinguish what one publishes as a civilian from what one publishes as a service member on active duty? Based on the personal experience of the SAF member interviewed for this paper, it is not, seeing as every time he posts anything on social media, his 'friends' regard this as being from the armed forces' perspective (Interview 3 2019). However, the use of social media for members of the armed forces is, in fact, limited due to OPSEC (Operational Security), meaning that operational data is not allowed to be shared with the public. Every member of the armed forces has to be informed as to what falls under OPSEC limitations, which are their main guidelines when posting and publishing online, even during work or while being deployed.

The second opinion is that it is "imperative to formulate a policy and guidelines for the use of social media platforms for business and private use when the SAF is

[11] All quotes from the interviews were translated into English by the authors of this paper.

involved" (Interview 2 2019). According to Nina Raduha, it is of "paramount importance for the day-to-day functioning of the SAF and its members, especially since learned and regulated behaviour on social media platforms in peace can transform into action in potential crisis situations and in missions abroad" (Interview 2 2019). Thus in addition to their primary purpose, rules are also necessary for several other reasons:

> Rules and policies are the basis for every action, they provide guidance and help and set limits, and they anticipate some consequences, which is why they are the cornerstone of any work – including the use of social networks. [...] They are necessary because they enable the safe and smart use of social media platforms (Interview 2 2019).

Of course, policies and regulations do not work magic and thus cannot eliminate all abusive and improper behaviour on social media platforms. Nevertheless, it is necessary to inform service members about the possible dangers and traps of social media, mainly from the point of view of protecting the armed forces and preserving data confidentiality.

Service members have to be particularly aware about the dangers and risks that the use of social media platforms brings, therefore educating them is imperative. An interesting and rather eye-opening experiment was recently conducted by NATO's Strategic Communications Centre of Excellence (NATO StratCom Coe), the goal of which was to "exploit social media and open source data to gather information on and influence military personnel during a military exercise" (Brown 2019).[12] The experiment underscores just how much personal information is free for the taking on social media, and – perhaps even more troubling – exactly how it can be used against even those of us who are the best positioned to resist it. Moreover, when professional service members are involved, this is even more dangerous. In fact, the results of the experiment "suggest that in the current digital arena an adversary would be able to collect enough personal data on soldiers to create targeted messages with precision, successfully influencing their chosen target audience to carry out desired behaviours" (Bay et al. 2019).

[12] Many of the details about how the operation worked remain classified, including precisely where it took place and which Allied force was involved. The NATO StratCom Coe group ran the drill during an exercise with military approval, however, service members were not aware of what was happening. Over the course of four weeks, the researchers developed fake pages and closed groups on Facebook that looked like they were associated with the military exercise, as well as profiles impersonating service members both real and imagined. The researchers also tracked down service members' Instagram and Twitter accounts and searched for other information available online, some of which an adversary might be able to exploit. By the end of the exercise, the researchers identified 150 soldiers, found the locations of several battalions, tracked troop movements, and compelled service members to engage in "undesirable behaviour", including leaving their positions against orders (Lapowsky 2019). More information on the experiment can be found in Bays 2019.

5 Conclusion: Dangers and Benefits of Social Media Use

Based on the literature review and the interviews, several benefits as well as several dangers regarding the use of social media by and in the armed forces can be identified. From an institutional perspective, the financial benefits of social media are a crucial aspect. Facebook, Instagram, and other social media platforms offer commercial and advertising opportunities for far less money than it would cost to hire an advertising agency and run a large-scale advertising campaign. Furthermore, social media platforms are an excellent way to reach the younger population, which is the main target demographic for recruitment endeavours. Social media platforms also enable two-sided interactive communication, meaning their use is not only about sharing information, but also about receiving feedback. Information is provided in real time, and communication is non-stop. However, this can also result in having inappropriate and insulting comments directed at you. Moreover, it demands quick responses from the institution communicating on social media. Seeing as information travels through social media in a split-second, this may pose a problem to the rigid and slow-responding armed forces.

Furthermore, the use of social media platforms enables networking and disseminating your own ideas and messages, making it very easy to connect with other groups sharing the same interests. In certain contexts, social media platforms also provide a mechanism for rapid command and control. Finally, for soldiers deployed to missions abroad, social media platforms offer opportunities to stay in touch with the home-front. It gives them the feeling of not being absent, although they actually are. Since social media platforms enable the fast sharing of pictures and audio files, the factor time is virtually non-present. However, as was already noted above, this is a double-edged sword, as too much information can cause the opposite effect. Instant interactive communication (conversations, photo sharing, messaging on Skype, Instagram or Facebook) with family members across an electronic divide – especially to and from war zones – can be simultaneously comforting and stressful. According to Dao (2011), "easy communication can relieve fears – but also stoke them. Once families become used to hearing from troops daily, lapses in communication can send imaginations racing."

The dangers of using social media platforms for the armed forces are actually the same as the dangers for civilians: fake news, identity theft, interpersonal bullying, leaked information, unofficial information prevailing over the official one, hate speech, and so forth. As one of the interviewees said, "the greatest danger of the digital age is preying upon the individual user, who might also be a member of the armed forces and who is not aware of the amount of information about themselves, their work, their loved ones, and their habits are available online" (Interview 2 2019). Soldiers deployed to missions abroad, to war zones, for instance, have to be aware of the dangers of posting their private information such as pictures of their family members. "You never know who may come to your homeland to try to seek revenge" (Interview 3 2019). Nevertheless, the use of social media on missions abroad can also be a great advantage, both in terms of private and professional use.

Yet it must be implemented professionally and in accordance with certain rules, which in some armed forces, such as the SAF, are still absent, and are therefore regulated by the contingent commander.

During recent years, social media platforms have proven that they have a vast potential and power to mobilise nations, to start revolutions, to influence elections. Therefore, it is imperative to educate state-leaders, professionals, and members of the armed forces not only about the possibilities, but also about the risks and dangers of social media. Social media platforms are used for a variety of organisational purposes, such as intelligence, information, and communication, as well as for personal purposes on the level of each individual member of the armed forces.

Armed forces which have acknowledged the existence and performance of social media in the age of modern security challenges and hybrid warfare use it as a tool of action and influence in order to reach their planned goals and fulfil their missions. Some of these countries, for instance the USA and the United Kingdom, are leading the way. They are extremely well regulated and highly proficient in directing the flow of activity on social networks as well as exploiting their full potential. Above all, these countries are well aware of the fact that limiting troops' activities on social media platforms is impossible, therefore it is better to coordinate and steer them. Offering education and training for the appropriate use of social media platforms is crucial in order for armed forces members to be highly proficient in using social media platforms safely, effectively, and efficiently both in a private and a professional environment.

Acknowledgments This article is based on the research done by the members of the Faculty of Social Sciences, University of Ljubljana.

References

Applewhite WL, Segal DR (1990) Telephone use by peacekeeping troops in the Sinai. Armed Forces Soc 17(1):117–126

Bartone VJ (2005) Missions alike and unlike: military families in peace and war. In: Presented at Inter-University Seminar on Armed Forces and Society, 45th Anniversary Biennial International Conference, Chicago, ZDA, 21–23 October 2005

Bay S, Betrolin G, Biteniece N, Christie EH, Dek A, Fredheim RE, Gallacher JD, Kononova K, Marchenko T (2019) NATO report: responding to cognitive security challenges. [Online]. [Viewed 25 August 2019]. Available from: https://www.stratcomcoe.org/responding-cognitive-security-challenges

BBC News (2019) Russia bans smartphones for soldiers over social media fears. [online]. [Viewed 25 August 2019]. Available from: https://www.bbc.com/news/world-europe-47302938

Bell DB, Schumm WR, Knott B, Ender MG (1999) The desert fax: a research note on calling home from Somalia. Armed Forces Soc 23(3):509–521

Brown R (2019) Researchers used fake social media accounts to influence NATO troops during military exercise. [Online]. [Viewed 25 August 2019]. Available from: https://edition.cnn.com/2019/02/20/politics/nato-social-media-influence-report/index.html

Chaffey D (2019) Global social media research summary 2019. [Online]. [Viewed 27 August 2019]. Available from: https://www.smartinsights.com/social-media-marketing/social-media-strategy/new-global-social-media-research/
Chalmers M (2011) Social media allow military families a deeper connection. USA Today, 24 November 2011. [Online]. [Viewed 26 August 2019]. Available from: http://usatoday30.usatoday.com/news/military/story/2011-11-28/military-deployment-social-media/51349158/1
Coser L (1974) Greedy institutions: patterns of undivided commitment. Free Press, New York
Dao J (2011) Staying in touch with home, for better or worse. The New York Times [online]. [Viewed 1 September 2019]. Available from: http://www.nytimes.com/2011/02/17/us/17soldiers.html
Ender MG (2009) American soldiers in Iraq. Routledge, New York
Gadgets Now (2020) Indian Army has a warning for its officers on using WhatsApp, Facebook: all you need to know. [Online]. [Viewed 20 January 2020]. Available from: https://www.gadgetsnow.com/slideshows/indian-army-has-a-warning-for-its-officers-on-using-whatsapp-facebook-15-things-to-know/Indian-Army-has-a-warning-for-its-officers-on-using-WhatsApp-Facebook-All-you-need-to-know/photolist/3058977.cms
Gupta A (2017) Social media: catalyst for empowering the armed forces. Scholar Warrior. [Online]. [Viewed 25 August 2019]. Available from: https://www.claws.in/images/journals_doc/593929489_11_chap.pdf
Hajjar RM (2014) Emergent postmodern US military culture. Armed Forces Soc 40(1):118–145
Hatch SL, Harvey SB, Dandeker C, Burdett H, Greenberg N, Fear NT, Wessely S (2013) Life in and after the armed forces: social networks and mental health in the UK military. Sociol Health Illn 35(7):1045–1064
Hern A (2018) Fitness tracking app Strava gives away location of secret US army bases. [Online]. [Viewed 25 August 2019]. Available from: https://www.theguardian.com/world/2018/jan/28/fitness-tracking-app-gives-away-location-of-secret-us-army-bases
Interview 1 (2019) Interview with personnel in charge of social media platforms "Postani vojak", Ljubljana, June 2019
Interview 2 (2019) Interview with Major Nina Raduha. Ljubljana, July 2019
Interview 3 (2019) Interview with a member of the SAF. Ljubljana, August 2019
Jones N, Baines P (2013) Losing control? Social Media and Military Influence. The RUSI Journal 158(1):72–78
Lapowsky I (2019) NATO group catfished soldiers to prove a point about privacy. [Online]. [Viewed 25 August 2019]. Available from: https://www.wired.com/story/nato-stratcom-catfished-soldiers-social-media/
Matthews-Juarez P, Juarez PD, Faulkner RT (2013) Social media and military families: a perspective. J Hum Behav Soc Environ 23(6):769–776
Moskos C (1986) Institutional/occupational trends in armed forces: an update. Armed Forces Soc 12(3):377–382
Nye JS Jr (1990) Soft power. Foreign Policy 80:153–171
Nye JS Jr (2004) Soft power: the means to success in world politics. PublicAffairs, New York
Paganini P (2013) Social media use in the military sector. [Online] InfoSec Institute, Elmwood Park. [Viewed 25 August 2019]. Available from: https://resources.infosecinstitute.com/social-media-use-in-the-military-sector/
Robb D (2017) Building the global heatmap. [Online]. [Viewed 25 August 2019]. Available from:. https://medium.com/strava-engineering/the-global-heatmap-now-6x-hotter-23fc01d301de
Segal MW (1986) The military and the family as greedy institutions. Armed Forces and Society Fall 13(1):9–38
Shachtman N (2009) Military blocks its own 'youtube' knockoff. [Online]. [Viewed 25 August 2019]. Available from: https://www.wired.com/2009/03/trooptube-block/
Svete U, Juvan J (2016) Soldiers' private digital communication as a disturbing factor. [Online]. Res Militaris, ISSN 2265-6294, 6(2). [Viewed 25 August 2019]. Available from: http://resmilitaris.net/index.php?ID=1024088
Wiatr J (1987) Sociologija vojske. Vojnoizdavački i novinski centar, Beograd

The manufacturer's authorised representative in the EU is Springer Nature Customer Service Centre GmbH, Europaplatz 3, 69115 Heidelberg, Germany. If you have any concerns regarding our products, please contact ProductSafety@springernature.com

Printed and bound by CPI Group (UK) Ltd, Croydon, CR0 4YY
30/09/2026
02225705-0003